# Lecture Notes in Computer Science 14894

Founding Editors

Gerhard Goos
Juris Hartmanis

Editorial Board Members

Elisa Bertino, *Purdue University, West Lafayette, IN, USA*
Wen Gao, *Peking University, Beijing, China*
Bernhard Steffen , *TU Dortmund University, Dortmund, Germany*
Moti Yung, *Columbia University, New York, NY, USA*

The series Lecture Notes in Computer Science (LNCS), including its subseries Lecture Notes in Artificial Intelligence (LNAI) and Lecture Notes in Bioinformatics (LNBI), has established itself as a medium for the publication of new developments in computer science and information technology research, teaching, and education.

LNCS enjoys close cooperation with the computer science R & D community, the series counts many renowned academics among its volume editors and paper authors, and collaborates with prestigious societies. Its mission is to serve this international community by providing an invaluable service, mainly focused on the publication of conference and workshop proceedings and postproceedings. LNCS commenced publication in 1973.

Djork-Arné Clevert · Michael Wand ·
Kristína Malinovská · Jürgen Schmidhuber ·
Igor V. Tetko
Editors

# AI in Drug Discovery

First International Workshop, AIDD 2024
Held in Conjunction with ICANN 2024
Lugano, Switzerland, September 19, 2024
Proceedings

 Springer

*Editors*
Djork-Arné Clevert 🆔
Machine Learning Research
Pfizer (Germany)
Berlin, Germany

Kristína Malinovská 🆔
Comenius University
Bratislava, Slovakia

Igor V. Tetko 🆔
Institute of Structural Biology, Molecular
Targets and Therapeutics Center
Helmholtz Zentrum München
Munich, Germany

Michael Wand 🆔
IDSIA USI-SUPSI
Viganello, Switzerland

Jürgen Schmidhuber 🆔
IDSIA USI-SUPSI
Viganello, Switzerland

ISSN 0302-9743          ISSN 1611-3349 (electronic)
Lecture Notes in Computer Science
ISBN 978-3-031-72380-3          ISBN 978-3-031-72381-0 (eBook)
https://doi.org/10.1007/978-3-031-72381-0

© The Editor(s) (if applicable) and The Author(s) 2025. This book is an open access publication.

**Open Access** This book is licensed under the terms of the Creative Commons Attribution 4.0 International License (http://creativecommons.org/licenses/by/4.0/), which permits use, sharing, adaptation, distribution and reproduction in any medium or format, as long as you give appropriate credit to the original author(s) and the source, provide a link to the Creative Commons license and indicate if changes were made.
The images or other third party material in this book are included in the book's Creative Commons license, unless indicated otherwise in a credit line to the material. If material is not included in the book's Creative Commons license and your intended use is not permitted by statutory regulation or exceeds the permitted use, you will need to obtain permission directly from the copyright holder.
The use of general descriptive names, registered names, trademarks, service marks, etc. in this publication does not imply, even in the absence of a specific statement, that such names are exempt from the relevant protective laws and regulations and therefore free for general use.
The publisher, the authors and the editors are safe to assume that the advice and information in this book are believed to be true and accurate at the date of publication. Neither the publisher nor the authors or the editors give a warranty, expressed or implied, with respect to the material contained herein or for any errors or omissions that may have been made. The publisher remains neutral with regard to jurisdictional claims in published maps and institutional affiliations.

This Springer imprint is published by the registered company Springer Nature Switzerland AG
The registered company address is: Gewerbestrasse 11, 6330 Cham, Switzerland

If disposing of this product, please recycle the paper.

# Preface

In recent years, Machine Learning has become more important than ever before. Large Language Models have revolutionized language-based tasks, with an impact far beyond the research community and IT-related industries: Artificial Intelligence for solving day-to-day tasks has become available for a wide range of end users across the world.

Machine Learning not only influences our daily lives, but also many fields of science and technology. As a specific example, we present Artificial Intelligence in organic chemistry and pharmaceutical research: a variety of tasks in this field are tackled with state-of-the-art Neural Network methods, leading to improved design and higher security of medical drugs, and to better solutions for chemical tasks in general, improving the quality of life of a large number of persons across the globe.

It is in this context that we proudly present the Proceedings of the 33th International Conference on Artificial Neural Networks (ICANN 2024). ICANN is the annual flagship conference of the European Neural Network Society (ENNS). This edition was co-organized by Istituto Dalle Molle di studi sull'intelligenza artificiale (IDSIA USI-SUPSI https://www.idsia.usi-supsi.ch) and by the Marie Skłodowska-Curie (MSC) Innovative Training Network European Industrial Doctorate "Advanced machine learning for Innovative Drug Discovery" (AIDD https://ai-dd.eu), supported by the MSC Doctoral Network "Explainable AI for Molecules" (AiChemist https://aichemist.eu). After two years of on-line and two years of hybrid conferences, ICANN 2024 was again organized as an in-person event, held on the premises of Università della Svizzera italiana (USI) and Scuola Universitaria Professionale della Svizzera italiana (SUPSI) in Lugano from September 17 to September 20, 2024.

ICANN 2024 featured three main conference tracks, namely Artificial Intelligence and Machine Learning, Bio-inspired Computing, and an Application Track. Dedicated members of the ICANN community also organized three workshops:

- AI in Drug Discovery
- Explainable AI in Human-Robot Interaction
- Reservoir Computing

    as well as three special sessions:

- Spiking Neural Networks and Neuromorphic Computing
- Accuracy, Stability, and Robustness in Deep Neural Networks
- Neurorobotics.

    Two tutorial sessions

- FEDn – A scalable federated machine learning framework for cross-device and cross-silo environments
- TSFEL - A Hands-on Introduction to Time Series Feature Extraction

were likewise proposed and organized by the community, as well as the

- Tox24 Challenge (prediction of toxicity of chemical compounds).

The proceedings of the conference are published as Springer volumes belonging to the Lecture Notes in Computer Science series. The conference had a total of 764 articles submitted to it. The papers went through a double-blind peer-review process supervised by experienced Area Chairs who suggested decisions to Program Chairs. In total, 564 Area Chairs, Program Committee (PC) members, and reviewers participated in the review process. The reviewers were on average assigned 3–4 articles each and submissions received on average 2.03 reviews each. A list of reviewers/PC Members who agreed to publish their names is included in the proceedings.

Based on the Area Chairs' and reviewers' comments, 310 articles (40.5% of initial submissions) were accepted, including 180 manuscripts selected for oral presentations. Out of the total number of accepted articles the majority (285 papers) were full articles with an average length of 15 pages, 20 manuscripts were short articles with an average length of 10 pages, and 5 were abstracts with an average length of 3 pages.

The accepted papers of the 33rd ICANN conference are published as 11 volumes, including one open-access volume with papers supported by the AIDD project.

The authors of accepted articles came from 29 different countries. As indicated by first author affiliation the largest number of articles came from China, followed by Germany, Japan, and Italy. While the majority of the articles were from academic researchers, the conference also attracted contributions from many industries including large pharmaceutical companies (Pfizer, Bayer, AstraZeneca, Johnson & Johnson), information and communication technology companies (Fujitsu and Baidu inc.), as well as multiple startups. This speaks to the increasing use of artificial neural networks in industry. Four keynote speakers were invited to give lectures on the timely aspects of advances in understanding the brain (Michael Reimann); new insights into cortical attention mechanisms and context-dependent gating and how they might inspire future developments in AI (Walter Senn); the current state of cognitive systems and how the full range of bio-signals can be utilized to further enhance human-robot interactions (Tanja Schultz); and a general overview of the past, present and future of machine learning (Jürgen Schmidhuber).

These proceedings provide comprehensive and up-to-date coverage of the dynamically developing field of Artificial Neural Networks. They are of major interest both for theoreticians as well as for applied scientists who are looking for new innovative approaches to solve their practical problems. We sincerely thank the Program and Steering Committee, Area Chairs, and the reviewers for their invaluable work.

September 2024

Djork-Arné Clevert
Michael Wand
Kristína Malinovská
Jürgen Schmidhuber
Igor V. Tetko

# Organization

## General Chairs

Jürgen Schmidhuber      KAUST Center of Generative AI, Saudi Arabia, and IDSIA USI-SUPSI, Switzerland

Igor V. Tetko      Helmholtz Munich, Germany and BigChem GmbH, Germany

## Program Chairs

Michael Wand      IDSIA USI-SUPSI, Switzerland and MeDiTech, SUPSI, Switzerland

Kristina Malinovska      Comenius University Bratislava, Slovakia

## Honorary Chair

Stefan Wermter      University of Hamburg, Germany

## Organizing Committee Chairs

Katya Ahmad      Helmholtz Munich, Germany

Alessandra Lintas      University of Lausanne, Switzerland

## Local Organizing Committee

Stefano van Gogh      IDSIA USI-SUPSI, Switzerland

Qinhan Hou      IDSIA USI-SUPSI, Switzerland

Nicolò La Porta      SUPSI, Switzerland

Alessandro Giusti      IDSIA USI-SUPSI, Switzerland

Vittorio Limongelli      USI, Switzerland

Cesare Alippi      IDSIA USI-SUPSI, Switzerland

Elena Invernizzi      IDSIA USI-SUPSI, Switzerland

Alessia Gianinazzi      IDSIA USI-SUPSI, Switzerland

## Communication Chairs

Sebastian Otte                University of Lübeck, Germany
R. Omar Chavez-Garcia         IDSIA USI-SUPSI, Switzerland

## Steering Committee

Stefan Wermter                University of Hamburg, Germany
Angelo Cangelosi              University of Manchester, UK
Igor Farkaš                   Comenius University Bratislava, Slovakia
Chrisina Jayne                Teesside University, UK
Matthias Kerzel               University of Hamburg, Germany
Alessandra Lintas             University of Lausanne, Switzerland
Kristína Malinovská           Comenius University Bratislava, Slovakia
Alessio Micheli               University of Pisa, Italy
Jaakko Peltonen               Tampere University, Finland
Brigitte Quenet               ESPCI Paris, France
Ausra Saudargiene             Lithuanian University of Health Sciences, and
                                 Vytautas Magnus University, Lithuania
Roseli Wedemann               Rio de Janeiro State University, Brazil
Sebastian Otte                University of Lübeck, Germany

## Area Chairs

Alessandro Antonucci          IDSIA USI-SUPSI, Switzerland
Alessandro Facchini           IDSIA USI-SUPSI, Switzerland
Alessio Micheli               University of Pisa, Italy
Anthony Cioppa                University of Liège, Belgium
Ausra Saudargiene             Lithuanian University of Health Sciences, and
                                 Vytautas Magnus University, Lithuania
Brigitte Quenet               ESPCI Paris PSL, France
Chen Zhao                     King Abdullah University of Science and
                                 Technology, Saudi Arabia
Daniele Palossi               IDSIA USI-SUPSI, Switzerland
Davide Bacciu                 University of Pisa, Italy
Fabio Rinaldi                 IDSIA USI-SUPSI, Switzerland
Felix Putze                   University of Bremen, Germany
Francesca Faraci              MeDiTech/BSP SUPSI-DTI, Switzerland
Gabriela Andrejková           P. J. Šafárik University in Košice, Slovakia
Hui Liu                       University of Bremen, Germany

| | |
|---|---|
| Igor Farkaš | Comenius University Bratislava, Slovakia |
| Kevin Jablonka | Friedrich Schiller University Jena, Germany |
| Marcello Restelli | Politecnico di Milano, Italy |
| Marco Forgione | IDSIA USI-SUPSI, Switzerland |
| Matthias Karlbauer | University of Tübingen, Germany |
| Michela Papandrea | ISIN, DTI, SUPSI, Switzerland |
| Mihai Andries | IMT Atlantique, France |
| Oleg Szehr | IDSIA USI-SUPSI, Switzerland |
| Rafael Cabañas de Paz | University of Almería, Spain |
| Silvio Giancola | King Abdullah University of Science and Technology, Saudi Arabia |
| Thang Vu | University of Stuttgart, Germany |
| Yibo Yang | King Abdullah University of Science and Technology, Saudi Arabia |
| Zuzana Černeková | Comenius University Bratislava, Slovakia |

## Workshop and Special Session Chairs

## Workshop: AI in Drug Discovery

| | |
|---|---|
| Djork-Arné Clevert | Pfizer GmbH, Germany |
| Igor Tetko | Helmholtz Munich, Germany |

## Workshop: Explainable AI in Human-Robot Interaction

| | |
|---|---|
| Stefan Wermter | University of Hamburg, Germany |
| Angelo Cangelosi | University of Manchester, UK |
| Igor Farkaš | Comenius University Bratislava, Slovakia |
| Theresa Pekarek-Rosin | University of Hamburg, Germany |

## Workshop: Reservoir Computing

| | |
|---|---|
| Alessio Micheli | University of Pisa, Italy |
| Gouhei Tanaka | Nagoya Institute of Technology, Japan |
| Claudio Gallicchio | University of Pisa, Italy |
| Benjamin Paassen | University of Bielefeld, Germany |
| Domenico Tortorella | University of Pisa, Italy |

## Special Session: Spiking Neural Networks and Neuromorphic Computing

Sander Bohté                    CWI Amsterdam, Netherlands
Sebastian Otte                  University of Lübeck, Germany

## Special Session: Accuracy, Stability, and Robustness in Deep Neural Networks

Vera Kurkova                    Institute of Computer Science of the Czech
                                    Academy of Sciences, Prague Czech Republic
Ivan Tyukin                     King's College, London, UK

## Special Session: Neurorobotics

Igor Farkaš                     Comenius University Bratislava, Slovakia
Kristína Malinovská             Comenius University Bratislava, Slovakia
Andrej Lúčny                    Comenius University Bratislava, Slovakia
Pavel Petrovič                  Comenius University Bratislava, Slovakia
Michal Vavrečka                 Czech Technical University in Prague, Czechia
Matthias Kerzel                 University of Hamburg, Germany
Hassan Ali                      University of Hamburg, Germany
Carlo Mazzola                   Italian Institute for Technology, Italy

## Program Committee

Abraham Yosipof                 CLB, Israel
Adam Arany                      KU Leuven, Belgium
Adrian Mirza                    Helmholtz Institute for Polymers in Energy
                                    Applications, Germany
Adrian Ulges                    RheinMain University of Applied Sciences,
                                    Germany
Alan Anis Lahoud                Örebro University, Sweden
Albert Weichselbraun            University of Applied Sciences of the Grisons
                                    (FHGR), Switzerland
Alessandra Roncaglioni          Istituto di Ricerche Farmacologiche Mario Negri,
                                    Italy
Alessandro Giusti               IDSIA USI-SUPSI, Switzerland
Alessandro Manenti              USI, Switzerland

| | |
|---|---|
| Alessandro Trenta | University of Pisa, Italy |
| Alessio Gravina | University of Pisa, Italy |
| Alex Doboli | Stony Brook University, USA |
| Alex Shenfield | Sheffield Hallam University, UK |
| Alexander Schulz | Bielefeld University, Germany |
| Alexandra Reichenbach | Heilbronn University of Applied Sciences, Germany |
| Ali Rodan | University of Jordan, Jordan |
| Alireza Raisiardali | Pragmatic Semiconductor Limited, UK |
| Aliza Subedi | Tribhuvan University, Nepal |
| Amir Mohammad Elahi | EPFL, Switzerland |
| Ana Claudia Sima | SIB Swiss Institute of Bioinformatics, Switzerland |
| Ana Sanchez-Fernandez | Johnson & Johnson Innovative Medicine, Belgium/JKU Linz, Austria |
| Andrea Licciardi | ICAR-CNR, Italy |
| Andreas Mayr | Johannes Kepler University Linz, Austria |
| Andreas Plesner | ETH Zurich, Switzerland |
| Andrej Lucny | Comenius University Bratislava, Slovakia |
| Aneri Muni | University of Montreal and Mila AI Institute, Canada |
| Angeliki Pantazi | IBM Research - Zurich, Switzerland |
| Angelo Moroncelli | IDSIA USI-SUPSI, Switzerland |
| Anne-Gwenn Bosser | Lab-STICC, ENIB, France |
| Anthony Strock | Stanford University, USA |
| Antonio Liotta | Free University of Bozen-Bolzano, Italy |
| Aparna Raj | BITS Pilani, Dubai Campus, United Arab Emirates |
| Ardian Selmonaj | IDSIA USI-SUPSI, Switzerland |
| Arnaud Gucciardi | University of Ljubljana, Slovenia |
| Artur Xarles | Universitat de Barcelona, Spain |
| Asma Sattar | University of Pisa, Italy |
| Aurelio Raffa Ugolini | Politecnico di Milano, Italy |
| Baohua Zhang | Beijing Institute of Technology, China |
| Baojin Huang | Wuhan University, China |
| Barbara Hammer | Bielefeld University, Germany |
| Bikram Kumar De | Texas State University, USA |
| Blerina Spahiu | University of Milan-Bicocca, Italy |
| Bo Li | Baidu Inc., China |
| Bogdan Kwolek | AGH University of Krakow, Poland |
| Bojian Yin | Innatera B.V., Netherlands |
| Brian Moser | German Research Center for Artificial Intelligence, Germany |

| | |
|---|---|
| Bulcsú Sándor | Babeş-Bolyai University, Romania |
| Cesare Donati | Politecnico di Torino, Italy |
| Chengeng Liu | Wuhan University, China |
| Chenxing Wang | Beijing University of Posts and Telecommunications, China |
| Chi Xie | Tongji University, China |
| Chong Zhang | Xi'an Jiaotong-Liverpool University, China |
| Chrisina Jayne | Teesside University, UK |
| Christoph Reinders | Leibniz University Hannover, Germany |
| Chrysoula Kosma | École Normale Supérieure Paris-Saclay, France |
| Cleber Zanchettin | Universidade Federal de Pernambuco, Brazil |
| Congcong Zhou | Sir Run Run Shaw Hospital, Zhejiang University, China |
| Coşku Can Horuz | University of Lübeck, Germany |
| Cunjian Chen | Monash University, Australia |
| Cyril Zakka | Stanford University, USA |
| Dania Humaidan | University Hospital Tübingen and Hertie Institute for Clinical Brain Research, Germany |
| Daniel Frank | University of Stuttgart, Germany |
| Daniel Nissani (Nissensohn) | Independent Research, Israel |
| Daniel Ortega | University of Stuttgart, Germany |
| Daniel Rose | University of Vienna, Austria |
| Daniele Angioletti | Università della Svizzera italiana, Switzerland |
| Daniele Castellana | Università degli Studi di Firenze, Italy |
| Daniele Malpetti | IDSIA USI-SUPSI, Switzerland |
| Daniele Zambon | IDSIA USI-SUPSI, Switzerland |
| Darío Ramos López | University of Almería, Spain |
| Davide Borra | University of Bologna, Italy |
| Dehui Kong | Sanechips; ZTE, China |
| Denis Kleyko | Örebro University, Sweden |
| Diana Borza | Babeş-Bolyai University, Romania |
| Dinesh Kumar | Bennett University, India |
| Dirk Väth | University of Stuttgart, Germany |
| Dongmian Zou | Duke Kunshan University, China |
| Doreen Jirak | Istituto Italiano di Tecnologia, Italy |
| Douglas McLelland | BrainChip, France |
| Duarte Folgado | Fraunhofer Portugal AICOS, Portugal |
| Dulani Meedeniya | University of Moratuwa, Sri Lanka |
| Dumitru-Clementin Cercel | Politehnica University of Bucharest, Romania |
| Dylan Muir | SynSense, Switzerland |
| Dylan R. Ashley | IDSIA USI-SUPSI, Switzerland |

E. J. Solteiro Pires            Universidade de Trás-os-Montes e Alto Douro,
                                  Portugal
Elena Šikudová                  Charles University, Czech Republic
Elia Cereda                     IDSIA USI-SUPSI, Switzerland
Elia Piccoli                    University of Pisa, Italy
Emma Svensson                   Johannes Kepler University, Austria and
                                  AstraZeneca, Sweden
Emmanuel Okafor                 King Fahd University of Petroleum and Minerals,
                                  Saudi Arabia
Evaldo Mendonça Fleury Curado   Centro Brasileiro de Pesquisas Físicas and
                                  National Institute of Science and Technology
                                  for Complex Systems, Brazil
Evgeny Mirkes                   University of Leicester, UK
Farhad Nooralahzadeh            Zurich University of Applied Sciences, University
                                  of Zurich, Switzerland
Fatemeh Hadaeghi                University Medical Center Hamburg-Eppendorf
                                  (UKE), Germany
Fatima Ezzeddine                Università della Svizzera italiana, Switzerland
Federico Errica                 NEC Laboratories Europe, Germany
Fedor Scholz                    University of Tübingen, Germany
Filipe Miguel Cardoso Micu      Helmholtz Munich, Germany
  Menezes
Flávio Arthur Oliveira Santos   Universidade Federal de Pernambuco, Brazil
Florian Lux                     University of Stuttgart, Germany
Francesco Faccio                IDSIA USI-SUPSI, Switzerland/KAUST AI
                                  Initiative, Saudi Arabia
Francesco Landolfi              Università di Pisa, Italy
Francis Colas                   Inria, France
Frédéric Alexandre              Inria, France
Gabriel Haddon-Hill             Keio University, Japan
Gabriela Sejnova                Czech Technical University in Prague, Czech
                                  Republic
Gabriele Lagani                 ISTI-CNR, Italy
Gerrit A. Ecke                  Mercedes-Benz AG, Germany
Gianvito Losapio                Politecnico di Milano, Italy
Giorgia Adorni                  IDSIA USI-SUPSI, Switzerland
Giovanni Dispoto                Politecnico di Milano, Italy
Giovanni Donghi                 University of Padua, Italy
Giuliana Monachino              University of Applied Sciences and Arts of
                                  Southern Switzerland (SUPSI), Switzerland
Gugulothu Narendhar             TCS Research, India
Guillaume Godin                 BigChem, Switzerland
Habib Irani                     Texas State University, USA

| | |
|---|---|
| Hanno Gottschalk | TU Berlin, Germany |
| Haoran Yang | Sichuan University, China |
| Hasby Fahrudin | AIBrain, South Korea |
| Hicham Boudlal | Mohammed First University of Oujda, Morocco |
| Hitesh Laxmichand Patel | Oracle/New York University, USA |
| Houssem Ouertatani | IRT SystemX & Univ. Lille, CNRS, Inria, France |
| Huang Yifan | Northeast Electric Power University, China |
| Hubert Cecotti | California State University, Fresno, USA |
| Hugo Cesar de Castro Carneiro | Universität Hamburg, Germany |
| Huifang Ma | Northwest Normal University, China |
| Igor Tetko | Helmholtz Munich, Germany |
| Ivor Uhliarik | Comenius University Bratislava, Slovakia |
| Jan Kalina | Czech Academy of Sciences, Institute of Computer Science, Czech Republic |
| Jan Niehues | KIT, Germany |
| Jan Prosi | University of Tübingen/International Max Planck Research School for Intelligent Systems, Germany |
| Jan Wollschläger | Bayer Pharmaceuticals, Germany |
| Jannis Vamvas | University of Zurich, Switzerland |
| Jérémie Cabessa | University of Versailles Saint-Quentin, France |
| Jia Cai | Guangdong University of Finance and Economics, China |
| Jiahui Chen | Xiamen University, China |
| Jialiang Xu | Soochow University, China |
| Jian Zhang | Zhejiang University, China |
| Jing Han | University of Cambridge, UK |
| Jingzehua Xu | Tsinghua University, China |
| Jinlai Ning | King's College London, UK |
| Jiong Wang | Beijing Normal University, China |
| Jiwen Yu | Peking University, China |
| Jizhe Yu | Dalian University of Technology, China |
| João Ricardo Sato | Universidade Federal do ABC, Brazil |
| Johannes Kriebel | University of Münster, Germany |
| Johannes Zierenberg | Max Planck Institute for Dynamics and Self-Organization, Germany |
| Jorge Lo Presti | University of Pavia, Italy |
| Julian Cremer | Pfizer, Germany |
| Julie Keisler | EDF R&D, Inria, France |
| Julien Marteen Akay | Bielefeld University of Applied Sciences and Arts, Germany |
| Jun Zhou | Wuhan University, China |

Junjie Zhou                    Nanjing University of Aeronautics and
                               Astronautics, China
Junzhou Chen                   College of William and Mary, USA
Kai Mao                        Xi'an Jiaotong University, China
Kevin Scheck                   University of Bremen, Germany
Keyan Jin                      Macao Polytechnic University, Macao SAR,
                               China
Khoa Phung                     University of the West of England, UK
Kiran Lekkala                  University of Southern California, USA
Kohei Nakajima                 University of Tokyo, Japan
Konstantinos Chatzilygeroudis  University of Patras, Greece
Krechel Dirk                   RheinMain University of Applied Science,
                               Germany
Kristína Malinovská            Comenius University Bratislava, Slovakia
Lapo Frascati                  ODYS, Italy
Laura Azzimonti                IDSIA USI-SUPSI, Switzerland
Laurent Larger                 FEMTO-ST Institute, Université
                               Bourgogne-Franche-Comté, France
Laurent Mertens                KU Leuven, Belgium
Laurent Udo Perrinet           Institut des Neurosciences de la Timone, Aix
                               Marseille Univ - CNRS, France
Lazaros Iliadis                Democritus University of Thrace, Greece
Lea Multerer                   IDSIA USI-SUPSI, Switzerland
Lei Li                         University of Copenhagen, Denmark
Lenka Tetkova                  Technical University of Denmark, Denmark
Leon Scharwächter              University of Tübingen, Germany
Leonardo Olivetti              Uppsala University, Sweden
Lewis Mervin                   AstraZeneca, UK
Lina Humbeck                   Boehringer Ingelheim Pharma GmbH & Co. KG,
                               Germany
Lindsey Vanderlyn              University of Stuttgart, Germany
Logofatu Doina                 Frankfurt University of Applied Sciences,
                               Germany
Lu Yang                        Wuhan University, China
Lubomir Antoni                 Pavol Jozef Šafárik University in Košice, Slovakia
Luca Butera                    IDSIA USI-SUPSI, Switzerland
Luca Sabbioni                  ML cube, Italy
Luís Gonçalves                 Universidade Federal de Pernambuco, Brazil
Lyra Puspa                     Vanaya NeuroLab, Indonesia and Canterbury
                               Christ Church University, UK
Maëlic Neau                    ENIB, France/Flinders University, Australia
Mahsa Abazari Kia              Northeastern University London, UK
Maksim Makarenko               Saudi Aramco, Saudi Arabia

| | |
|---|---|
| Manas Mejari | IDSIA USI-SUPSI, Switzerland |
| Manon Dampfhoffer | Univ. Grenoble Alpes, CEA, List, France |
| Manuel Traub | University of Tübingen, Germany |
| Marco Paul E. Apolinario | Purdue University, USA |
| Marco Podda | University of Pisa, Italy |
| Marco Tarabini | Politecnico di Milano, Italy |
| Marcondes Ricarte da Silva Júnior | Federal University of Pernambuco, Brazil |
| Marek Suppa | Comenius University Bratislava, Slovakia |
| Marina Garcia de Lomana | Bayer AG, Germany |
| Markus Heinonen | Aalto University, Finland |
| Marta Lenatti | Consiglio Nazionale delle Ricerche, Italy |
| Martin Lefebvre | Université catholique de Louvain, Belgium |
| Martin Ritzert | Georg-August Universität Göttingen, Germany |
| Masanobu Inubushi | Tokyo University of Science, Japan |
| Matej Fandl | Comenius University Bratislava, Slovakia |
| Matej Pecháč | Tachyum s.r.o., Slovakia |
| Matteo Rufolo | IDSIA USI-SUPSI, Switzerland |
| Matthias Kerzel | Universität Hamburg, Germany |
| Matthias Rupp | Luxembourg Institute of Science and Technology, Luxembourg |
| Matus Tuna | Comenius University Bratislava, Slovakia |
| Maximilian Kimmich | University of Stuttgart, Germany |
| Maynara Donato de Souza | Federal University of Pernambuco, Brazil |
| Mengdi Li | University of Hamburg, Germany |
| Mengjia Zhu | IMT School for Advanced Studies Lucca, Italy |
| Michal Bechny | UNIBE/SUPSI, Switzerland |
| Michal Burgunder | Università della Svizzera italiana, Switzerland |
| Michal Vavrecka | CIIRC CTU, Czech Republic |
| Michela Sperti | Politecnico di Torino, Italy |
| Michele Fontanesi | University of Pisa, Italy |
| Mikhail Andronov | Università della Svizzera Italiana, Switzerland |
| Mingyang Li | Stanford University, USA |
| Mingyong Li | Chongqing Normal University, China |
| Miroslav Strupl | IDSIA USI-SUPSI, Switzerland |
| Moritz Wolter | Rheinische Friedrich-Wilhelms-Universität Bonn, Germany |
| Muhammad Arslan Masood | Aalto University, Finland |
| Muhammad Burhan Hafez | University of Southampton, UK |
| Mykhailo Sakevych | Texas State University, USA |
| Nabeel Khalid | German Research Center for Artificial Intelligence, Germany |
| Navdeep Singh Bedi | IDSIA USI-SUPSI, Switzerland |

| | |
|---|---|
| Nicolò La Porta | Università della Svizzera Italiana, Switzerland |
| Niklas Beuter | Technische Hochschule Lübeck, Germany |
| Niko Dalla Noce | University of Pisa, Italy |
| Oh-hyeon Choung | dsm-firmenich, Switzerland |
| Olivier J. M. Béquignon | Leiden University, The Netherlands |
| Omran Ayoub | University of Applied Sciences and Arts of Southern Switzerland, Switzerland |
| Oscar Mendez Lucio | Recursion, Spain |
| Osvaldo Simeone | King's College London, UK |
| Otto Brinkhaus | Spleenlab GmbH, Germany |
| Pascal Tilli | University of Stuttgart, Germany |
| Paul Czodrowski | JGU Mainz, Germany |
| Paul Kainen | Georgetown University, USA |
| Paula Štancelová | Comenius University Bratislava, Slovakia |
| Paula Torren-Peraire | Johnson & Johnson Innovative Medicine, Belgium |
| Pavel Denisov | University of Stuttgart, Germany |
| Pavel Kordík | Czech Technical University in Prague, Czech Republic |
| Pavel Petrovič | Comenius University Bratislava, Slovakia |
| Peiyu Liang | Temple University, USA |
| Peng Qiao | NUDT, China |
| Pengjie Liu | Southern University of Science and Technology, China |
| Pengyu Li | Yanshan University, China |
| Peter Hartog | Helmholtz Munich, Germany and AstraZeneca, Sweden |
| Petia Koprinkova-Hristova | Institute of Information and Communication Technologies, Bulgarian Academy of Sciences, Bulgaria |
| Petra Vidnerová | Institute of Computer Science, Czech Academy of Sciences, Czech Republic |
| Philipp Allgeuer | University of Hamburg, Germany |
| Plinio Moreno | Instituto Superior Técnico/University of Lisbon, Portugal |
| Qinhan Hou | IDSIA USI-SUPSI, Switzerland |
| Quentin Jodelet | Tokyo Institute of Technology, Japan |
| Raphael Yokoingawa de Camargo | Universidade Federal do ABC, Brazil |
| Răzvan-Alexandru Smădu | National University of Science and Technology POLITEHNICA Bucharest, Romania |
| Reyan Ahmed | University of Arizona, USA |
| Ricardo O. Chávez García | IDSIA USI-SUPSI, Switzerland |
| Riccardo Massidda | Università di Pisa, Italy |

| | |
|---|---|
| Riccardo Renzulli | University of Turin, Italy |
| Robert Legenstein | Graz University of Technology, Austria |
| Robertas Damaševičius | Kaunas University of Technology, Lithuania |
| Robin Winter | Pfizer, Germany |
| Rodolphe Vuilleumier | École normale supérieure-PSL, Sorbonne Université, CNRS, France |
| Rodrigo Braga | NOVA School of Science and Technology, Portugal |
| Rodrigo Clemente Thom de Souza | Federal University of Paraná, Brazil |
| Roseli S. Wedemann | Universidade do Estado do Rio de Janeiro, Brazil |
| Roxane Jacob | University of Vienna, Austria |
| Ru Zhou | RuiJin Hospital LuWan Branch, Shanghai Jiaotong University School of Medicine, China |
| Ruinan Wang | University of Bristol, UK |
| Ruixi Zhou | Beijing University of Posts and Telecommunications, China |
| Rupesh Raj Karn | New York University Abu Dhabi, United Arab Emirates |
| Samuel Genheden | AstraZeneca R&D, Sweden |
| Sandra Mitrovic | IDSIA USI-SUPSI, Switzerland |
| Sankalp Jain | NCATS-NIH, USA |
| Sara Joubbi | University of Pisa, Italy |
| Seema Dilipkumar Aswani | BITS Pilani, Dubai Campus, UAE |
| Seiya Satoh | Tokyo Denki University, Japan |
| Semih Beycimen | Cranfield University, UK |
| Senhui Qiu | Ulster University, UK |
| Sergei Katkov | Free University of Bozen-Bolzano, Italy |
| Sergio Mauricio Vanegas Arias | LUT University, Finland |
| Shangchao Su | Fudan University, China |
| Sheng Xu | Chinese University of Hong Kong, Shenzhen, China |
| Shenyang Liu | University of Central Florida, USA |
| Sherjeel Shabih | Humboldt University, Germany |
| Shi Haoran | China Water Northeastern Investigation, Design & Research Co., Ltd., China |
| Shingo Murata | Keio University, Japan |
| Shinnosuke Matsuo | Kyushu University, Japan |
| Shiyao Zhang | University of Bremen, Germany |
| Sho Shirasaka | Osaka University, Japan |
| Simiao Zhuang | TUM Beijing, China |
| Simon Heilig | Ruhr University Bochum, Germany |
| Šimon Horvát | University of Ljubljana, Slovakia |
| Simone Bonechi | University of Siena, Italy |

| | |
|---|---|
| Simone Lionetti | Hochschule Luzern, Switzerland |
| Siyu Wu | Central South University of Forests and Technology, China |
| Stefano Damato | IDSIA USI-SUPSI, Switzerland |
| Stéphane Meystre | MeDiTech/SUPSI, Switzerland |
| Steve Azzolin | University of Trento, Italy |
| Sudip Roy | Indian Institute of Technology Roorkee, India |
| Sujala D. Shetty | BITS Pilani, Dubai Campus, United Arab Emirates |
| Taoran Fu | Hunan University & Hunan Institute of Engineering, China |
| Teste Olivier | Université Toulouse 2, IRIT (UMR5505), France |
| Thierry Viéville | Inria, France |
| Tianyi Wang | Nanyang Technological University, Singapore |
| Tim Schlippe | IU International University of Applied Sciences, Germany |
| Tingyu Lin | TU Wien, Austria |
| Tuan Le | Pfizer, Germany |
| Valerie Vaquet | Bielefeld University, Germany |
| Vangelis Metsis | Texas State University, USA |
| Vani Kanjirangat | IDSIA USI-SUPSI, Switzerland |
| Varun Ojha | Newcastle University, UK |
| Veronica Lachi | Fondazione Bruno Kessler, Italy |
| Viktor Kocur | Comenius University Bratislava, Slovakia |
| Vincenzo Palmacci | University of Vienna, Austria |
| Wei Dai | Robo Space, China |
| Weiqi Li | Peking University, China |
| Weiran Chen | Soochow University, China |
| Wenjie Zhang | Shandong University, China |
| Wenwei Gu | Chinese University of Hong Kong, China |
| Wolfram Schenck | Bielefeld University of Applied Sciences and Arts, Germany |
| Xavier Hinaut | Inria, France |
| Xi Wang | National University of Defense Technology, China |
| Xiangxian Li | Shandong University, China |
| Xiangyuan Peng | Technical University of Munich, Germany |
| Xiaochen Yuan | Macao Polytechnic University, Macao SAR, China |
| Xiaomeng Fu | University of Chinese Academy of Sciences, China |
| Xiaowen Sun | University of Hamburg, Germany |
| Xiaoxiao Miao | Singapore Institute of Technology, Singapore |

| | |
|---|---|
| Xingda Yao | Zhejiang University of Technology, China |
| Xinxin Luo | Southeast University, China |
| XinZhi Lin | Beihang University, China |
| Xun Lin | Beihang University, China |
| Yan Jiang | Nanjing University of Information Science and Technology, China |
| Yang Cao | Shanghai University of Finance and Economics, China |
| Yangfan Zhou | Southwest University of Science and Technology, China |
| Yangxun Ou | East China Normal University, China |
| Yao Du | Beihang University, China |
| Yaxin Hu | University of Lübeck, Germany |
| Ye Hu | Pfizer, Germany |
| Yi Li | Lancaster University, UK |
| Yichi Zhang | Fudan University, China |
| Yiming Tang | Shanghai Lixin University of Accounting and Finance, China |
| Ying Tan | Key Laboratory for Computer Systems of State Ethnic Affairs Commission, Southwest Minzu University, China |
| Yiqing Shen | Johns Hopkins University, USA |
| Yixuan Xiao | University of Stuttgart, Germany |
| Yong Luo | Wuhan University, China |
| Yongtao Tang | National University of Defense Technology, China |
| Yuankun Chen | University of Science and Technology, China |
| Yuansheng Ma | Soochow University, China |
| Yuchen Guo | Institute of Information Engineering, Chinese Academy of Sciences, China |
| Yuichi Katori | Future University Hakodate, Japan |
| Yuji Kawai | Osaka University, Japan |
| Yusen Wu | Sichuan University, China |
| Yutaka Nakamura | Riken, Japan |
| Yuya Okadome | Tokyo University of Science, Japan |
| Zdravko Marinov | Karlsruhe Institute of Technology, Germany |
| Zeyao Liu | Key Institute of Information Engineering, Chinese Academy of Sciences, China |
| Zhang Ke | China University of Petroleum (Beijing), China |
| Zhenjie Yao | Institute of Microelectronics of the Chinese Academy of Sciences, China |
| Zheyan Gao | Tianjin University, China |
| Zhiheng Qiu | City University of Macau, China |

Zhihuan Xing                    Beihang University, China
Zuzana Berger Haladova          Comenius University Bratislava, Slovakia

# Plenary Talks

# Past, Present, Future, and Far Future of Machine Learning

Jürgen Schmidhuber

IDSIA USI-SUPSI, Switzerland, and KAUST AI Initiative, Saudi Arabia

I'll discuss modern Artificial Intelligence and how the principles of the G, P and T in Chat GPT emerged in 1991. I'll also discuss what's next in AI, and its expected impact on the future of the universe.

# Dendritic Computations and Deep Learning in the Brain

Walter Senn

University of Bern, Institut für Physiologie, Computational Neuroscience Lab,
Switzerland

Artificial Intelligence, through its working horse of neural networks, is inspired by the biological example of the brain. The unprecedented success of AI in modeling cognitive processes, in turn, inspires functional models of the brain. Yet, when looking into the brain, additional biological structures become apparent, such as dendritic morphologies, interneuron circuits, recurrent connectivity, error representations, top-down signaling and various gating hierarchies. I will give a review on these biological elements and show how they may integrate in an energy-based theory of cortical computation. Dendrites and cortical microcircuits turn out to implement a real-time version of error-backpropagation based on prospective errors. The theory is inspired by the least-action principle in physics from which all dynamical equations of motions are derived. We likewise derive the neuronal dynamics, including the synaptic dynamics with gradient-descent learning, from our Neuronal Least-Action (NLA) principle. The principle tells that the cortical activities and the real-time learning follows a path that minimizes prospective errors across all neurons of the network. Prospective errors in output neurons relate to behavioral errors, while prospective errors in deep network neurons relate to errors in the neuron-specific dendritic prediction of somatic firing. I will explain how these ideas relate to cortical attention mechanisms and context-dependent gating that link to, and potentially inspires, recent developments in AI.

# Biosignal-Adaptive Cognitive Systems

Tanja Schultz

University of Bremen, Fachbereich 3 - Mathematik und Informatik, Cognitive Systems Lab, Germany

I will describe technical cognitive systems that automatically adapt to users' needs by interpreting their biosignals: Human behavior includes physical, mental, and social actions that emit a range of biosignals which can be captured by a variety of sensors. The processing and interpretation of such biosignals provides an inside perspective on human physical and mental activities, complementing the traditional approach of merely observing human behavior. As great strides have been made in recent years in integrating sensor technologies into ubiquitous devices and in machine learning methods for processing and learning from data, I argue that the time has come to harness the full spectrum of biosignals to understand user needs. I will present illustrative cases ranging from silent and imagined speech interfaces that convert myographic and neural signals directly into audible speech, to interpretation of human attention and decision making in human-robot interaction from multimodal biosignals.

# Workshop: AI in Drug Discovery

The dramatic increase in the use of Artificial Intelligence (AI) and traditional machine learning methods in different scientific fields has become an essential asset in the future development of the chemical industry, including the pharmaceutical, agro biotech, and other chemical sectors. The Workshop on AI in Drug Discovery collected cutting-edge contributions in the rapidly evolving field of AI-driven drug discovery. Submissions encompassing various facets such as generative models, explainable AI, model distillation, uncertainty quantification, reaction informatics and synthetic route prediction, quantum machine learning for reactivity, methodologies for mining very large compound data sets, federated learning, analysis of HTS data and identification of frequent hitters and other topics related to the use of ML in chemistry were considered. In total 12 submissions were selected for oral talks and 12 articles were presented as posters.

The covered topics included:

- Big Data and advanced machine learning in chemistry
- eXplainable AI (XAI) in chemistry
- Chemoinformatics
- Use of deep learning to predict molecular properties
- Modeling and prediction of chemical reaction data
- Generative models

As part of the workshop, the Tox24 Challenge [1] was organized in collaboration with the Chemical Research in Toxicology journal and the AIDD https://ai-dd.eu and AiChemist https://aichemist.eu projects. The training and test sets consisted of chemicals and compounds that have been tested for activity against Transthyretin (TTR) by the US Environmental Protection Agency. Participants competed for a prize of 1000 €, which was awarded to the developers of the winning model during the closing ceremony.

The authors of articles/abstracts of the AIDD workshop were invited to submit their articles to the special issue of J. Cheminformatics.

## Keynote

Artem Cherkasov             University of British Columbia, Vancouver, Canada

## Organizers

Djork-Arné Clevert          Pfizer GmbH, Germany
Igor Tetko                  Helmholtz Munich
Katya Ahmad                 Helmholtz Munich, Germany

# Workshop Website

https://ai-dd.eu/icann2024

# Reference

1. Tetko, I. V.: Tox24 Challenge. Chem. Res. Toxicol. **37**(6), 825–826 (2024). https://doi.org/10.1021/acs.chemrestox.4c00192.

# The Use of Active Learning for Effective Exploration of the Chemical Universe

Ekaterina Manskaia🅳 and Artem Cherkasov🅳

Vancouver Prostate Centre, University of British Columbia, Vancouver, BC, V6T 1Z4 Canada

artc@mail.ubc.ca

Over recent decades, drug discovery has heavily relied on in-silico methods of hits identification, since the use of CADD can significantly minimize the costs and time of the process [1–3]. The major CADD tool, molecular docking, emerged as a computationally efficient alternative to resource-intensive wet lab screening of up to million-molecules-sized chemical databases [4].

In recent years chemical libraries have dramatically expanded reaching the levels beyond dozens of billions of entities [5]. Since traditional docking relies on a brute-force computation, such a tremendous database increase not only improved the chances of finding effective hits [6, 7] but also forced in-silico methods to adapt to the reality of new Big Data and to appropriate elements of machine learning (ML), artificial intelligence (AI) and active learning (AL) [8–10].

A number of ML-accelerated docking techniques have recently evolved. An early example of AL integrated with ML methods was Progressive Docking, which aimed to emulate docking scores for a subset of a database and develop quantitative structure-activity relationship (QSAR) models integrated into active learning (AL) iterative cycles [16]. In 2020, we released the Deep Docking (DD), an AI-driven platform enhanced with AL [17]. DD facilitated efficient exploration of billions of molecules and consistently provided reliable access to extensive docking results using moderate computational resources [17]. DD has demonstrated significant success in various studies, achieving up to a 6000-fold enrichment for top-ranked hits and accelerating screening processes by up to 50 times compared to brute-force methods [17].

Following the launch of DD, several research groups have revisited the AL concept in docking including MEMES [20], MolPAL [21], AutoQSAR/DeepChem [22, 23], and the approach developed by Xu et al. [24], among many others.

While AL has proven indispensable for exploring large datasets, it has also found successful applications in diverse domains, including lead optimization and drug response prediction models [25-30].

To summarize, the recent use of ML methods for the exploration of the Chemical Universe has revolutionized all phases of drug discovery. Among those methods, Active Learning methodologies have proven to be particularly effective in significantly reducing computational demands for ultra-large screening campaigns. Continued research aimed at enhancing active learning techniques will maximize the potential of chemical space exploration, thereby accelerating the discovery of novel therapeutics.

# References

1. Keserü, G.M., Makara, G.M.: The influence of lead discovery strategies on the properties of drug candidates. Nat. Rev. Drug Discovery **8**, 203–212 (2009)
2. Tropsha, A., Isayev, O., Varnek, A., et al.: Integrating QSAR modelling and deep learning in drug discovery: the emergence of deep QSAR. Nat. Rev. Drug Discovery **23**, 141–155 (2024). https://doi.org/10.1038/s41573-023-00832-0
3. Jorgensen, W.L.: The many roles of computation in drug discovery. Science **303**, 1813–1818 (2004)
4. Murgueitio, M.S., Rakers, C., Frank, A., et al.: Balancing inflammation: computational design of small-molecule toll-like receptor modulators. Trends Pharmacol. Sci. **38**(2), 155–168 (2017)
5. Warr, W.A., Nicklaus, M.C., Nicolaou, C.A., et al.: Exploration of ultralarge compound collections for drug discovery. J. Chem. Inf. Model. **62**(9), 2021–2034 (2022)
6. Cartblanche22 Homepage. https://cartblanche.docking.org/. Accessed 12 June 2023
7. Potlitz, F., Link, A., Schulig, L.: Advances in the discovery of new chemotypes through ultra-large library docking. Expert Opin. Drug Discov. **18**(3), 303–313 (2023)
8. Popova, M., Isayev, O., Tropsha, A.: Deep reinforcement learning for de novo drug design. Sci. Adv. **4**(7), 1–14 (2018)
9. Pu, L., Naderi, M., Liu, T., et al.: EToxPred: a machine learning-based approach to estimate the toxicity of drug candidates. BMC Pharmacol. Toxicol. **20**(2), 1–15 (2019)
10. Van Vleet, T.R., Liguori, M.J., Lynch, J.J., et al.: Screening strategies and methods for better off-target liability prediction and identification of small-molecule pharmaceuticals. SLAS Discov. **24**(1), 1–24 (2019)
11. Yoon, S., Smellie, A., Hartsough, D., et al.: Surrogate docking: structure-based virtual screening at high throughput speed. J. Comput. Aided Mol. Des. **19**(7), 483–497 (2005)
12. Naik, A.W., Kangas, J.D., Sullivan, D.P., et al.: Active machine learning-driven experimentation to determine compound effects on protein patterns. eLife **5**, 1–21 (2016)
13. Kuan, J., Radaeva, M., Avenido, A., et al.: Keeping pace with the explosive growth of chemical libraries with structure-based virtual screening. WIREs Comput. Mol. Sci. **13**(6), 1–15 (2023)
14. Ren, P., Xiao, Y., Chang, X., et al.: A survey of deep active learning. ACM Comput. Surv. **54**(9), 1–40 (2022)
15. Vasanthakumari, P., Zhu, Y., Brettin, T., et al.: A comprehensive investigation of active learning strategies for conducting anti-cancer drug screening. Cancers **16**(3), 530, 1–18 (2024)

16. Cherkasov, A., Ban, F., Li, Y., et al.: Progressive docking: a hybrid QSAR/docking approach for accelerating in silico high throughput screening. J. Med. Chem. **49**(25), 7466–7478 (2006)

17. Gentile, F., Agrawal, V., Hsing, M., et al.: Deep docking: a deep learning platform for augmentation of structure-based drug discovery. ACS Central Sci. **6**(6), 939–949 (2020)

18. Ton, A-T., Gentile, F., Hsing, M., et al.: Rapid identification of potential inhibitors of SARS-CoV-2 main protease by deep docking of 1.3 billion compounds. Mol. Inf. **39**(8), 1–18 (2020)

19. Rossetti, G.G., Ossorio, M.A., Rempel, S., et al.: Non-covalent SARS-CoV-2 M$^{pro}$inhibitors developed from in silico screen hits. Sci. Rep. **12**, 2505 (2022). https://doi.org/10.1038/s41598-022-06306-4

20. Mehta, S., Laghuvarapu, S., Pathak, Y., et al.: MEMES: machine learning framework for enhanced MolEcular screening. Chem. Sci. **12**(35), 11710–11721 (2021)

21. Graff, D.E., Shakhnovich, E.I., Coley, C.W.: Accelerating high-throughput virtual screening through molecular pool-based active learning. Chem. Sci. **12**(22), 7866–7881 (2021)

22. Lyu, J., Wang, S., Balius, T.E., et al.: Ultra-large library docking for discovering new chemotypes. Nature **566**(7743), 224–229 (2019)

23. Yang, Y., Yao, K., Repasky, M.P., et al.: Efficient exploration of chemical space with docking and deep learning. J. Chem. Theory Comput. **17**(11), 7106–7119 (2021)

24. Xu, Y.Y., Wang, Q., Xu, G.Q., et al.: Active learning-driven discovery and dynamics simulation of novel SQLE inhibitors with ADMET analysis and molecular modification. Preprint available at Research Square (2024). https://doi.org/10.21203/rs.3.rs-4163089/v1

25. Thompson, J., Walters, W.P., Feng, J.A., et al.: Optimizing active learning for free energy calculations. Artif. Intell. Life Sci. **2**, 100050, 1–11 (2022)

26. Gusev, F., Gutkin, E., Kurnikova, M.G., et al.: Active learning guided drug design: Lead optimization based on relative binding free energy modeling. J. Chem. Inf. Model. **63**(2), 583–594 (2023)

27. Khalak, Y., Tresadern, G., Hahn, D.F., et al.: Chemical space exploration with active learning and alchemical free energies. J. Chem. Theory Comput. **18**(10), 6259–6270 (2022)

28. Konze, K.D., Bos, P.H., Dahlgren, M.K., et al.: Reaction-based enumeration, active learning, and free energy calculations to rapidly explore synthetically tractable chemical space and optimize potency of cyclin-dependent Kinase 2 inhibitors. J. Chem. Inf. Model. **59**(9), 3782–3793 (2019)

29. Zaverkin, V., Holzmüller, D., Steinwart, I., et al.: Exploring chemical and conformational spaces by batch mode deep active learning. Digital Discov. **1**, 605–620 (2022)

30. Tynes, M., Gao, W., Burrill, D.J., et al.: Pairwise Difference regression: a machine learning teta-algorithm for improved prediction and uncertainty quantification in chemical search. J. Chem. Inf. Model. **61**(8), 3846–3857 (2021)

31. Jastrzębski, S., Szymczak, M., Pocha, A., et al.: Emulating docking results using a deep neural network: a new perspective for virtual screening. J. Chem. Inf. Model. **60**(9), 4246–4262 (2020)

32. Martin, L.: State of the art iterative docking with logistic regression and Morgan fingerprints. ChemRxiv (2021). Preprint accessible at https://doi.org/10.26434/che mrxiv.14348117.v1

33. Berenger, F., Kumar, A., Zhang, K.Y.J., et al.: Lean-docking: exploiting ligands' predicted docking scores to accelerate molecular docking. J. Chem. Inf. Model. **61**(5), 2341–2352 (2021)

34. Kalliokoski, T.: Machine learning boosted docking (HASTEN): an open-source tool to accelerate structure-based virtual screening campaigns. Mol. Inf. **40**(9), 2100089, 1–5 (2021)

35. Choi, J.; Lee, J.: V-Dock: fast generation of novel drug-like molecules using machine-learning-based docking score and molecular optimization. Int. J. Mol. Sci. **22**(21), 11635 (2021)

36. Morris, C.J., Stern, J.A., Stark, B., et al.: MILCDock: machine learning enhanced consensus docking for virtual screening in drug discovery. J. Chem. Inf. Model. **62**(22), 5342–5350 (2022)

37. García-Ortegón, M., Simm, G.N.C., Tripp, A.J., et al.: DOCKSTRING: easy molecular docking yields better benchmarks for ligand design. J. Chem. Inf. Model. **62**(15), 3486–3502 (2022)

# Contents

**Abstracts from the AIDD Workshop**

# Enhancing Interpretability in Molecular Property Prediction with Contextual Explanations of Molecular Graphical Depictions

Marco Bertolini[1,2]($\boxtimes$) ⓘ, Linlin Zhao[3] ⓘ, Floriane Montanari[1] ⓘ, and Djork-Arné Clevert[1,2] ⓘ

[1] Machine Learning Research, Bayer AG, 13353 Berlin, Germany
[2] Machine Learning Research, Pfizer Worldwide Research Development and Medical, Berlin, Germany
`marco.bertolini@pfizer.com`
[3] Field Solutions, Bayer AG, 40789 Monheim am Rhein, Germany

**Abstract.** The field of explainable AI applied to molecular property prediction models has often been reduced to deriving atomic contributions. This has impaired the interpretability of such models, as chemists rather think in terms of larger, chemically meaningful structures, which often do not simply reduce to the sum of their atomic constituents. We develop an explanatory strategy yielding both local as well as more complex structural attributions. We derive such contextual explanations in pixel space, exploiting the property that a molecule is not merely encoded through a collection of atoms and bonds, as is the case for string- or graph-based approaches. We provide evidence that the proposed explanation method satisfies desirable properties, namely sparsity and invariance with respect to the molecule's symmetries, to a larger degree that the SMILES-based counterpart model. Nonetheless, they correlate as expected with these string-based explanation as well as with ground truths, when available. Contextual explanations thus maintain the accuracy of the original explanations while improving their interpretability.

## 1 Introduction

The rapid development of Deep Learning (DL) models for molecular property prediction [9,23,25] has increased the need for equally powerful interpretability methods. These are crucial to gain trust in the model, understand its limitations, and support the chemist's knowledge and intuition in the process of property optimization. An ideal explainable AI (XAI) framework for molecular property prediction would assign attributions to both individual atoms and larger substructures. Additionally, it would also be able to provide ideas of modifications that can be made to the structure to overcome a particular issue.

Common modeling strategies involve fully connected networks from precomputed molecular fingerprints [19] or latent representations and, when enough

© The Author(s) 2025
D.-A. Clevert et al. (Eds.): AIDD 2024, LNCS 14894, pp. 1–12, 2025.
https://doi.org/10.1007/978-3-031-72381-0_1

training data is available, end-to-end training with graph convolutional networks (GCNs) [7,17]. Explanations for these types of models can take the form of atomic attributions (particularly for GCNs [10,24]), or feature importance using packages such as SHAP [15]. Many of our in-house models are built upon the CDDD embedding space [22], which poses a challenge for explainability. The CDDD space is the bottleneck layer of a pre-trained autoencoder translating between different SMILES representations of molecules. One approach to explainability consists of assigning the attributions to the original SMILES, i.e., tracing back gradients through the pre-trained encoder. However, the interpretation and visualization of attributions for string characters is challenging [11]. Additionally, the validity of the use of gradients for discrete character inputs can also be questioned [1].

In this work, we propose a novel XAI approach tailored to networks built upon CDDD descriptors. This method, which we refer to as *contextual explainability*, is able to capture both atomic and structural contributions. We rely on explainability of concepts derived in the context of image analysis [2–4,8,12,18,21] as well as on Img2Mol, a recently published optical molecular recognition model [6], that is able to translate images of molecules to their CDDD embeddings. We find that early layers in Img2Mol capture basic chemical features like atoms and bonds, while deeper layers learn more complex chemical structures, for instance, rings. By aggregating explanations from all the layers [5], we show that we can provide sparse and robust explanations that respect molecular symmetry and show both very localized highlights for particular atoms and more global importance for entire substructures. Moreover we also provide evidence that our contextual explanations are faithful, that is, they agree with ground-thruth ones, when available. We also show that they agree, as expected, with the character-based explanations obtained through the original CDDD encoder.

## 2   Setup

Our explainability framework relies on the recently proposed Img2Mol model. It consists of a convolutional neural network whose task is to map molecular graphical depictions to their CDDD embeddings. The CDDD space $\mathcal{C} = [-1, 1]^{512}$ is constructed as the bottleneck layer of a Seq2Seq-autoencoder network trained to translate several million chemically-equivalent SMILES representations of molecules and defines a continuous molecular descriptor, which can be utilized as a powerful input for training downstream tasks. Figure 1a depicts the structure of the Img2Mol encoder. Img2Mol is trained on over ten million unique canonical SMILES and establishes the new state-of-the-art performance in reconstruction accuracy. The training objective consists in minimizing the distance in CDDD space between the Img2Mol embeddings and the embeddings obtained through the encoder from [22]. The reconstruction from CDDD to SMILES to evaluate the model's performance occurs through the pre-trained decoder from [22]. For further details concerning the model architecture, as well as the training procedure and the model performance, we refer the reader to [6].

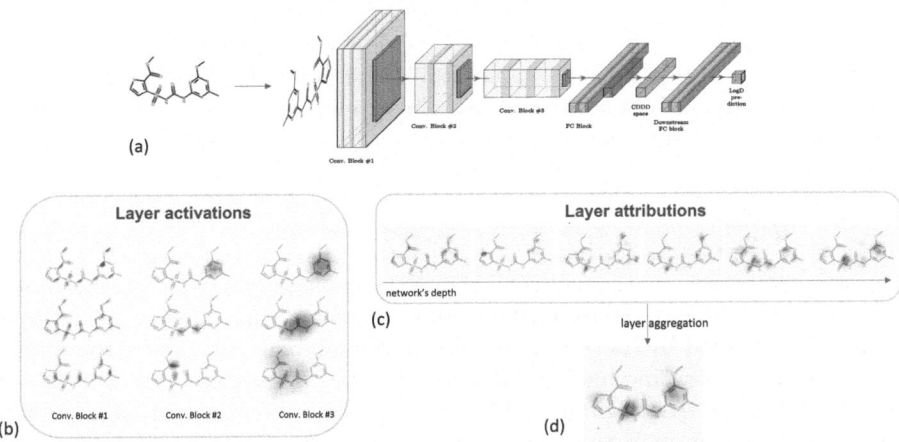

**Fig. 1.** Summary of the contextual explanation framework for molecular property predictions: (a) Architecture of the image-based QSAR model; (b) channel layer activations learned by Img2Mol; (c) layer attributions; (d) contextual explanation obtained by aggregating over the various layer attributions. Green (pink) overlay indicates positive (negative) contribution towards the prediction. (Color figure online)

We trained a quantitative structure-property relationship (QSAR) model to predict the lipophilicity of small molecules. The dataset consists of ~63000 molecules with measured values in a in-house logD assay. Specifically, the downstream model is a multilayer perceptron (MLP) with two hidden layers and has been trained on the molecules' CDDD embeddings. The model performance is excellent with a cluster cross-validation coefficient of determination ($R^2$) score of 0.902. Upon testing on an independent dataset of 62 molecules, whose endpoints have been reviewed and curated from the Pesticide Properties Database [14], the final model led to an $R^2$ score of 0.914. All the XAI experiments and examples presented in this work are obtained from this final model, where the CDDD embeddings are generated through the Img2Mol encoder network. All the used input molecules are obtained from public data.

We will also compare our explanations with those obtained through the original pretrained SMILES-based encoder. For these atom-based explanations, we follow the strategy of [26], which we briefly summarize here. The approach for assessing the impact of individual characters within a SMILES string consists in computing a character-wise sensitivity score, which is determined by averaging the predictions of the network $\Phi$ when a particular position is substituted with any character from the SMILES vocabulary. Specifically, for a character at position $i$ in a SMILES string $\mathbf{s}$ of length $n$, its contribution to the prediction $\Phi(\mathbf{s})$ is defined as

$$A^i = \Phi(\mathbf{s}) - \frac{1}{|\nu|} \sum_{k \in \nu} \Phi(\tilde{\mathbf{s}}^i_k) \, , \qquad (1)$$

where $i = 1, \ldots, n$, $\nu$ is the set of all possible characters in the SMILES vocabulary, and $\tilde{\mathbf{s}}^i_k$ represents the modified SMILES string with the character at position $i$ replaced by character $k$ from $\nu$.

## 3   Methods

Our strategy is based on the fact that 1) deep layers in neural networks learn high-level concepts and 2) for pure convolutional networks, the value of each "super-pixel" is determined by its receptive field in input space [16]. We combine these two properties by tracing back attributions to pixel space through the Img2Mol encoder instead of the original CDDD encoder. We remark that this is possible since both encoders map the respective inputs to the CDDD space. Explicitly, let $\Lambda : \mathcal{C} \to \mathbb{R}$ be the QSAR downstream model and Img2Mol $: \mathcal{M} \to \mathcal{C}$, where $\mathcal{M} \simeq [0, 255]^{224 \times 224}$ is the input space consisting of images of $224 \times 224$ pixels. We then construct the network $\Phi = $ Img2Mol $\circ \Lambda : \mathcal{M} \xrightarrow{\psi_p} \mathcal{M}_p \xrightarrow{\xi_p} \mathcal{C} \xrightarrow{\Lambda} \mathbb{R}$ by concatenating the Img2Mol encoder with the logD downstream network described in the previous section. Here, $\mathcal{M}_p$ is the output space of the $p^{\text{th}}$ convolutional layer in the network. $\mathcal{M}_p$ has dimension $k_p \times k_p \times C_p$, where $C_p$ is the number of channels in the $p^{\text{th}}$ layer, and $k_p$ is the size length of the embedding in terms of superpixels. Thus, our contextual explanations are obtained via the network $\Phi$ applied to a graphical depiction of the sample molecule.

Figure 1b shows a few channels activations, further grouped by the corresponding convolutional block. This example supports our intuition: while filters in early layers reduce to node, angle, and edge detectors, filters in deeper layers are activated by larger sub-structures in the molecule, e.g., rings and functional groups. It is then natural to use these layer attributions as a chemically meaningful feature basis for our explanations. Thus, we compute feature attributions values for each convolutional layer of the network $\Phi$, choosing gradients to measure feature importance. Explicitly, for each convolutional layer $p$ we compute superpixel attributions as

$$a_p(\mathbf{x}) = \sum_{c_p=1}^{C_p} \frac{\partial (\xi_{p,c_p} \circ \Lambda)(\mathbf{x})}{\partial \psi_{p,c_p}(\mathbf{x})} \times \psi_{p,c_p}(\mathbf{x}) \, , \qquad (2)$$

where the sum is over the channel dimension. The above formula formalizes our intuition: for a given convolutional layer $p$, each channel output activation $\psi_{p,c_p}(\mathbf{x})$ is weighted by its contribution $\partial (\xi_{p,c_p} \circ \Lambda)(\mathbf{x})/\partial \psi_{p,c_p}(\mathbf{x})$ to the endpoint prediction. The attribution method (2) is known as activation$\times$gradient, which is a natural extension of input$\times$gradient [20] to obtain layer-wise attributions. Our implementation of (2) is based on the **captum** package [13]. Figure 1c depicts some layer-wise explanations. We notice that attributions for early layers, as expected, focus on simple geometric features like atoms and bonds, in contrast

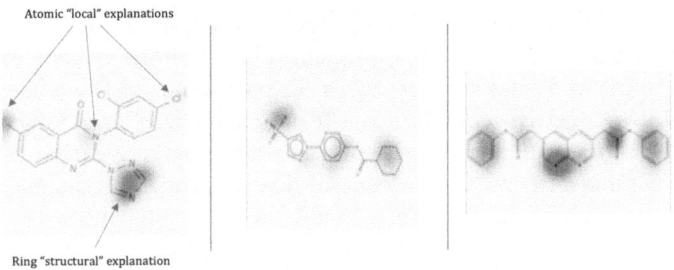

**Fig. 2.** Explanations are sparse and incorporate both local and structural features

to attributions for deeper layers, where explanations involve entire functional groups.

Finally, as we assume that an exhaustive explanation would involve a combination of both local and structural features, we propose a simple procedure to extract a single explanation from the layer attribution maps. Namely, we aggregate the maps (2) over all the convolutional layers to obtain a unique network-wide attribution map

$$a(\mathbf{x}) = \sum_{p \in \{\text{conv.layers}\}} a_p(\mathbf{x}) \ . \tag{3}$$

The above equation determines the weighting of the various local and structural components, as determined by the relative value of the different layer attributions, resulting in the final contextual attribution map. Figure 1d illustrates an example of the result of such an aggregation strategy.

## 4    Experiments and Properties of Contextual Explanations

In this section, we turn to examine some of the desired properties that our contextual explanations (3) possess, namely, sparsity, the interplay of both local and structural features, and the invariance of explanations with respect to the symmetries of the molecule's graphical depiction Fig. 2. We will then conclude by showing that the contextual explanations exhibit a strong correlation with the character-based explanations (1) and, in a simple setting, to known ground truths. Therefore, our proposed strategy generates explanations that are accurate, robust, and faithful to the underlying ground truth. These explanations are more interpretable due to their emphasis on chemical sub-structures, aligning more closely to a chemist's perspective, compared to the somewhat artificial approach of atom-based attributions. The molecules used in the experiments described below are publicly available compounds and were not included in the training or test sets.

**Contextual Explanations and Sparsity.** The examples in Fig. 4a illustrates the defining characteristic of our approach. The attribution heat map incorporates both atomic and structural features. In particular, in the left example, the model attributes positive contributions (indicated by a green overlay) to the outermost Cl atom and methyl group, while it assigns negative contributions (marked by a pink overlay) to the central N atom and the triazine ring. These assignments are in alignment with a medicinal chemist's intuition about logD contributions.

The aggregation procedure (3) has, in addition, a denoising effect. As can be seen in Fig. 1d, the aggregated map is more sparse than the individual layer attribution maps, as it concentrates only on the most important features contributing to the prediction. Sparsity is a desirable property for an explanation, as feature cluttering impairs the interpretability of the predictions.

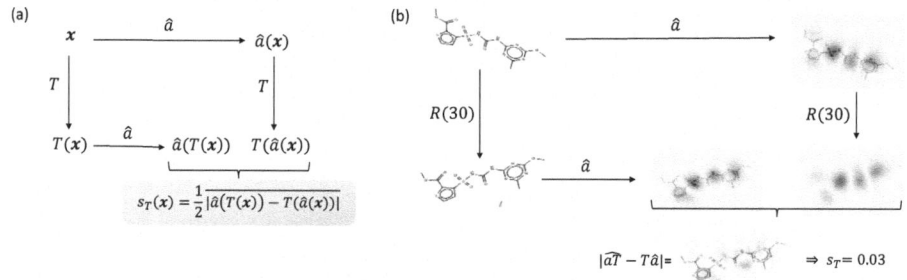

**Fig. 3.** Definition of the symmetry score (a) and example for a 30° rotation (b).

**Invariance with Respect to Molecule's Symmetries.** An important property for interpretability is that the explanations respect the symmetries of the input molecule. Among the CDDD-based methods, those based on SMILES will fail to produce invariant explanations, as the SMILES string representations explicitly break the molecule's symmetries. In what follows, we provide evidence that our contextual explanations, instead, tend to be invariant under such symmetries. Explicitly, let $\mathcal{T}$ be the symmetry group of a molecule's graphical depiction, that is, the group of image transformations that leave the chemical content invariant: given a molecule image $\mathbf{x}$ and a transformation $T \in \mathcal{T}$, then $\mathbf{x}' = T(\mathbf{x})$ corresponds to the same molecule. To quantify the invariance of our contextual explanations with respect to a symmetry group $\mathcal{T}$, we define the symmetry score for the transformation $T \in \mathcal{T}$ as

$$s_T(\mathbf{x}) = \frac{1}{2}\overline{|\widehat{a}(T(\mathbf{x})) - T(\widehat{a}(\mathbf{x}))|} \, , \tag{4}$$

where $\widehat{a}$ is obtained from (3) upon normalization to the range $[-1, 1]$, and the overline denotes the average in pixel space. The score is graphically illustrated in Fig. 3 with an example for a rotation of the molecular depiction of a 30° angle. The score measures the average absolute difference between two attribution maps, and thus provides a quantitative measure of the correlation between two explanations. In performing this average, we only include normalized attributions $\widehat{a}(\mathbf{x})$ in absolute value above a given threshold (0.05), to avoid the score to depend on the amount of white space in the picture. The score is normalized such that it takes value between 0 (which occurs when the transformation commutes with the attribution maps, $\widehat{a}T = T\widehat{a}$) and 1 (which occurs when $\widehat{a}T = -T\widehat{a}$ and $\widehat{a} = \pm 1$).

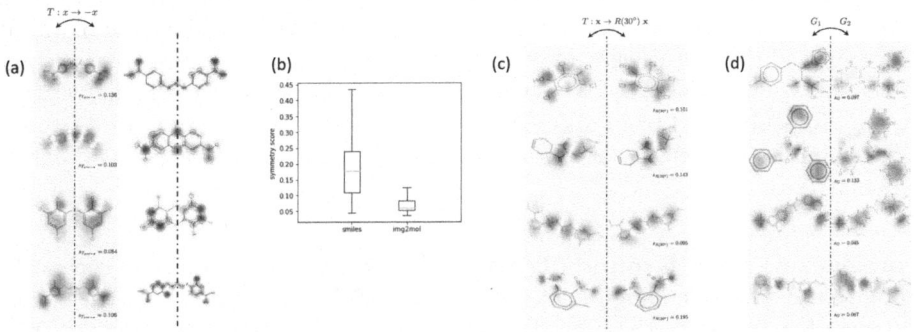

**Fig. 4.** Properties of contextual explainability for molecular property prediction. Explanations tend to preserve the molecule's depiction symmetry under (a) reflections and (c) rotations; (b) the symmetry score for reflection transformation is in average much lower for contextual explanations that for SMILES-based ones, showing a higher degree of symmetry invariance (lower is better); (d) explanations are robust with respect to different pictorial representations. Green (pink) overlay indicates positive (negative) contribution towards the prediction.

We compute the score (4) for two transformations, namely reflection across the vertical axis $T = T_{x \leftrightarrow -x}$, and rotation of a 30° angle in the plane of the image $T = R(30°)$. For reflections we report a value of $\mathbb{E}[s_{T_{x \leftrightarrow -x}}] = 0.135 \pm 0.003$, computed by averaging scores for 21 images of molecules exhibiting such symmetry. This value indicates that the symmetry is well captured by our explanations, as can be seen in Fig. 4b. For rotations we instead report an average score over 121 molecule images of $\mathbb{E}[s_{R(30°)}] = 0.169 \pm 0.004$, which again indicates that upon rotations, the attribution maps show a high consistency. We report in Fig. 4a-c some examples of such transformations, the respective explanations and the associated scores. Such examples provide a visual intuition that for the achieved values of the score, the symmetries are well-respected by our explanations. We note that these tend to be less sparse than the original contextual explanations due to the normalization we introduced in (4).

While not feasible for rotation transformations, we tested the symmetry invariance properties under reflection of the SMILES based explanations (1).

We reported the corresponding examples in Fig. 4a, and the scores for both contextual and character-based attributions in Fig. 4b. This quantitative analysis indicates that, on average, our contextual explanations respect the molecules' symmetry more faithfully than those derived from SMILES-based explanations.

**Robustness with Respect to the Graphical Depiction.** There are several standards for graphically representing a molecule structure. We provide evidence that our contextual explanations are robust with respect to different graphical representations by slightly modifying the score (4). Let $G_1(m), G_2(m)$ be two different graphical depictions of the same molecule m, then the score $s_G(m) = \frac{1}{2}|\widehat{a}(G_1(m)) - \widehat{a}(G_2(m))|$ measures the average absolute difference between two attribution maps obtained from the two different graphical methods. We computed the score across a set of 121 molecules, and we obtained an average value of $\mathbb{E}[s_G] = 0.148 \pm 0.003$, which reveals a high level of agreement between explanations obtained from different graphical representations. Figure 4d shows some examples of such pictorial representations with their respective contextual explanations.

**Correlation with Ground Truth.** Explainability methods are primarily employed to assess the alignment between a model's predictions and the salient features that influence these predictions, at least when they are accurate. This process is particularly daunting in the realm of chemistry, where it is often unclear even to experts which molecular structures contribute to a compound's specific properties. Due to this we design a simple task, where the ground truths are well-established, to assess whether our model's contextual explanations are consistent with known ground truths. Our experiment consists in training a classifier on the CDDD representations of molecules to detect the presence of benzene rings. As expected, our model achieves perfect accuracy on both training and test datasets. We then quantified the agreement between both the contextual and the character-based explanation with the ground truths, given by the atoms forming a benzine ring. We employ an overlap score, similar to the above, defined as $s_O(x) = \frac{1}{2}|\widehat{a}(x) - g(x)|$, where $g$ are the ground truths maps obtained by graphically depicting the attribution maps that assign value 1 to the carbon atoms of any benzine rings, and zero to all other atoms (there are no negative contributions to this task). The comparison then happens in pixel space. We observed that both the SMILES-based and the contextual explanations exhibit very similar results, with an average test score of about $\sim 0.15$ (Fig. 5), which signifies a strong agreement. In fact, we depicted as a benchmark the theoretical value $s_O = 0.4$ of the overlap score for random maps where the attributions are drawn from a uniform distribution $\mathcal{U}(0,1)$. We also computed the score $s_O$ for contextual explanations using an untrained classifier, and we observe that it still performs better than random maps. This is a confirmation of the fact that, while the classifier cannot solve the tasks, the attributions (2) are a combination of meaningful activations, among which benzine rings, which then tend to appear more often into the explanations than in randomly drawn maps. Thus, our contextual explanations have some intrinsically knowledge of chemistry structures, which help the interpretability of the explanations.

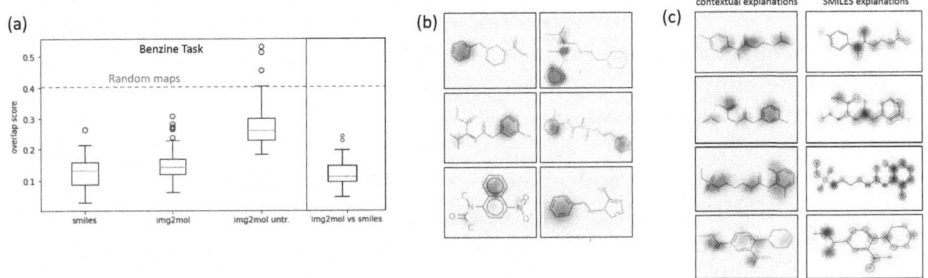

**Fig. 5.** Correlation between the contextual explanations and (a,b) ground truth and (a,c) SMILES-based attributions (lower is better).

**Correlation with SMILES-Based Explanations.** To investigate the relationship between contextual explanations and atomistic SMILES-based explanations (1), we overlay graphical depictions derived from SMILES explanations onto similar molecular representations. By analyzing the red and green image channels, we extract a map that quantifies the negative and positive attributions, respectively, facilitating direct comparison between the two explanation types. Examples in Fig. 5c illustrate the alignment between these explanations, showing a notable qualitative agreement. Additionally, we quantify their correlation using the overlap score defined above, $s_O(\mathbf{x}) = \frac{1}{2}\overline{|\widehat{a}(\mathbf{x}) - \widehat{a}_{\text{SM}}(\mathbf{x})|}$, where $\widehat{a}_{\text{SM}}$ denotes the graphical representation of the SMILES explanations. We assess a strong correlation, evidenced by a score of $s_O \sim 0.1$ (Fig. 5a on the right). This indicates an even more robust correlation than previously noted between the SMILES explanations and the ground truth. Such findings strongly support the reliability of our contextual explanations and the well-behaved structure of the CDDD space.

## 5  Conclusions

This work introduced an approach to explaining molecular property predictions based on molecules' graphical depictions, which we named contextual explainability. Our method is able to capture both basic (like atoms and bonds) as well as more complex structures (like rings and chemical groups), yielding explanations that are more aligned with chemists' intuition. We provided evidence that our contextual explanations possess several desirable properties: the attributions tend to be sparse, are robust with respect to the chosen graphical representation, and respect the symmetry of the input image. In addition, the strongly correlate with the corresponding SMILES-based attributions, obtained trough the pretrained CDDD encoder, as well as with known ground truths.

It would be interesting to explore our explanation framework in the context of property optimization: explanations in pixel space have the advantage that the model can explain a prediction not exclusively in terms of what is present in the given molecule, but also in terms of what is missing. The explanations could then be employed to provide suggestions of structure modifications for optimizing the given molecular property.

More generally, this approach is a first step into designing explainability frameworks leveraging the multi-modality of the data. In this case, one modality (string-based SMILES) is suitable to train a highly informative embedding to train downstream models, while the other (pixel space) yields more inherently interpretable results.

# References

1. Akita, H., et al.: BayesGrad: explaining predictions of graph convolutional networks. In: International Conference on Neural Information Processing, pp. 81–92. Springer (2018). https://doi.org/10.1007/978-3-030-04221-9_8
2. Alain, G., Bengio, Y.: Understanding intermediate layers using linear classifier probes. arXiv preprint arXiv:1610.01644 (2016)
3. Bau, D., Zhou, B., Khosla, A., Oliva, A., Torralba, A.: Network dissection: quantifying interpretability of deep visual representations. In: Proceedings of the IEEE Conference on Computer Vision and Pattern Recognition, pp. 6541–6549 (2017)
4. Bertolini, M., Clevert, D.A., Montanari, F.: Explaining, evaluating and enhancing neural networks' learned representations. In: Iliadis, L., Papaleonidas, A., Angelov, P., Jayne, C. (eds.) Artificial Neural Networks and Machine Learning - ICANN 2023, pp. 269–287. Springer Nature Switzerland, Cham (2023). https://doi.org/10.1007/978-3-031-44192-9_22
5. Bertolini, M., et al.: From slides (through tiles) to pixels: an explainability framework for weakly supervised models in pre-clinical pathology (2023)
6. Clevert, D.A., Le, T., Winter, R., Montanari, F.: Img2Mol - accurate SMILES recognition from molecular graphical depictions. Chem. Sci. **12**(42), 14174–14181 (2021). https://doi.org/10.1039/D1SC01839F
7. Duvenaud, D.K., et al.: Convolutional networks on graphs for learning molecular fingerprints. In: Cortes, C., Lawrence, N.D., Lee, D.D., Sugiyama, M., Garnett, R. (eds.) Advances in Neural Information Processing Systems 28, pp. 2224–2232. Curran Associates, Inc. (2015), http://papers.nips.cc/paper/5954-convolutional-networks-on-graphs-for-learning-molecular-fingerprints.pdf
8. Engel, J., Hoffman, M., Roberts, A.: Latent constraints: learning to generate conditionally from unconditional generative models. arXiv preprint arXiv:1711.05772 (2017)
9. Gilmer, J., Schoenholz, S.S., Riley, P.F., Vinyals, O., Dahl, G.E.: Neural message passing for quantum chemistry. In: International Conference on Machine Learning, vol. 70, pp. 1263–1272 (06–11 Aug 2017). http://proceedings.mlr.press/v70/gilmer17a.html

10. Henderson, R., Clevert, D.A., Montanari, F.: Improving molecular graph neural network explainability with orthonormalization and induced sparsity. In: Proceedings of the 38th International Conference on Machine Learning, pp. 4203–4213 (2021)

11. Karpov, P., Godin, G., Tetko, I.V.: Transformer-CNN: swiss knife for QSAR modeling and interpretation. J. Cheminformatics **12**, 1–12 (2020)

12. Kim, B., et al.: Interpretability beyond feature attribution: quantitative testing with concept activation vectors (TCAV). In: International Conference on Machine Learning, pp. 2668–2677. PMLR (2018)

13. Kokhlikyan, N., et al.: Captum: A unified and generic model interpretability library for PyTorch (2020)

14. Lewis, K.A., Tzilivakis, J., Warner, D.J., Green, A.: An international database for pesticide risk assessments and management. Hum. Ecol. Risk Assess. Int. J. **22**(4), 1050–1064 (2016)

15. Lundberg, S.M., Lee, S.I.: A unified approach to interpreting model predictions. In: Guyon, I., et al. (eds.) Advances in Neural Information Processing Systems 30, pp. 4765–4774. Curran Associates, Inc. (2017). http://papers.nips.cc/paper/7062-a-unified-approach-to-interpreting-model-predictions.pdf

16. Luo, W., Li, Y., Urtasun, R., Zemel, R.: Understanding the effective receptive field in deep convolutional neural networks. In: Proceedings of the 30th International Conference on Neural Information Processing Systems, pp. 4905–4913 (2016)

17. Montanari, F., Kuhnke, L., Ter Laak, A., Clevert, D.A.: Modeling physico-chemical ADMET endpoints with multitask graph convolutional networks. Molecules **25**(1) (2020)

18. Raghu, M., Poole, B., Kleinberg, J., Ganguli, S., Sohl-Dickstein, J.: On the expressive power of deep neural networks. In: International Conference on Machine Learning, pp. 2847–2854. PMLR (2017)

19. Rogers, D., Hahn, M.: Extended-connectivity fingerprints. J. Chem. Inf. Model. **50**(5), 742–754 (2010). https://doi.org/10.1021/ci100050t, pMID: 20426451

20. Shrikumar, A., Greenside, P., Shcherbina, A., Kundaje, A.: Not just a black box: learning important features through propagating activation differences (2017)

21. Szegedy, C., Vanhoucke, V., Ioffe, S., Shlens, J., Wojna, Z.: Rethinking the inception architecture for computer vision. In: Proceedings of the IEEE Conference on Computer Vision and Pattern Recognition, pp. 2818–2826 (2016)

22. Winter, R., Montanari, F., Noé, F., Clevert, D.A.: Learning continuous and data-driven molecular descriptors by translating equivalent chemical representations. Chem. Sci. **10**, 1692–1701 (2019). https://doi.org/10.1039/C8SC04175J

23. Wu, Z., et al.: MoleculeNet: a benchmark for molecular machine learning. Chem. Sci. **9**, 513–530 (2018). https://doi.org/10.1039/C7SC02664A

24. Xie, S., Lu, M.: Interpreting and understanding graph convolutional neural network using gradient-based attribution method (2019)

25. Yang, K., et al.: Analyzing learned molecular representations for property prediction. J. Chem. Inf. Model. **59**(8), 3370–3388 (2019). https://doi.org/10.1021/acs.jcim.9b00237, pMID: 31361484

26. Zhao, L., Montanari, F., Heberle, H., Schmidt, S.: Modeling bioconcentration factors in fish with explainable deep learning. Artificial Intelligence in the Life Sciences **2**, 100047 (2022). https://doi.org/10.1016/j.ailsci.2022.100047, https://www.sciencedirect.com/science/article/pii/S2667318522000174

**Open Access** This chapter is licensed under the terms of the Creative Commons Attribution 4.0 International License (http://creativecommons.org/licenses/by/4.0/), which permits use, sharing, adaptation, distribution and reproduction in any medium or format, as long as you give appropriate credit to the original author(s) and the source, provide a link to the Creative Commons license and indicate if changes were made.

The images or other third party material in this chapter are included in the chapter's Creative Commons license, unless indicated otherwise in a credit line to the material. If material is not included in the chapter's Creative Commons license and your intended use is not permitted by statutory regulation or exceeds the permitted use, you will need to obtain permission directly from the copyright holder.

# Temporal Evaluation of Probability Calibration with Experimental Errors

Hannah Rosa Friesacher[1,3]($\boxtimes$) (ID), Emma Svensson[2,3](ID), Adam Arany[1](ID),
Lewis Mervin[4](ID), and Ola Engkvist[3,5](ID)

[1] ESAT-STADIUS, KU Leuven, Leuven 3000, Belgium
[2] ELLIS Unit Linz, Institute for Machine Learning, Johannes Kepler University Linz,
Linz 4040, Austria
[3] Molecular AI, Discovery Sciences, R&D, AstraZeneca Gothenburg, Mölndal 431 83,
Sweden
rosafriesacher@live.at
[4] Molecular AI, Discovery Sciences, R&D, AstraZeneca Cambridge, Cambridge CB2
0AA, UK
[5] Department of Computer Science and Engineering, Chalmers University of
Technology, Gothenburg 412 96, Sweden

## 1 Introduction

The quantification of uncertainties associated with neural network predictions can facilitate optimal decision-making and accelerate workflows where time and resource efficiency are essential. In drug discovery, computational tools exist that estimate predictive uncertainties to enable the assessment of costs and risk in the discovery and development pipeline [11]. There are various sources of uncertainty in machine learning. A common classification found in literature is the distinction between aleatoric uncertainty, which originates from uncertainty in the data, and epistemic sources, which quantifies uncertainty inherent in the choice of model. We refer to Hüllermeier & Waegemann [6] and Gruber et al. [3] for a deeper discussion of uncertainty sources. It is important to point out that modern neural networks often fail to give realistic estimates of the uncertainty associated with a prediction in classification tasks, resulting in poorly calibrated models [4,11]. There are various calibration methods for classification models, that aim to obtain better uncertainty estimates by fitting a calibrating model to a separate dataset in a post-hoc manner. Another strategy to achieve more reliable predictions is the incorporation of model uncertainty, by taking into account model variance, which increases when the model is overfitting or the test instance lies outside the domain of the training data. This work compares the performance of single-task classification models trained on industry-scale assay data in a temporal analysis. In contrast to random or cluster-based strategies to split the data, temporal splits simulate most accurately the drug discovery pipeline in pharmaceutical companies [16]. A temporal splitting strategy enables model training on older data and prediction on subsequent folds. We use temporal splits to compare the performance and calibration of Random Forest (RF)

© The Author(s) 2025
D.-A. Clevert et al. (Eds.): AIDD 2024, LNCS 14894, pp. 13–20, 2025.
https://doi.org/10.1007/978-3-031-72381-0_2

models for classification tasks with and without post-hoc calibration using two different calibration approaches. Furthermore, we investigate whether the inclusion of data uncertainty in the form of probabilistic labels improves uncertainty estimation. Finally, we use the temporal setting to investigate how the temporal evolution of the test set affects model calibration.

## 2    Methods

We evaluate single-task classification models on data from ten assays and two assay categories, including 'Panel' and 'Other' assays [5]. The assays are labeled using the assay category combined with a number from 1 to 5, e.g. 'Panel-1'. The 'Panel' category comprises cross-project assays such as undesired off-target effects, whereas 'Other' includes project-specific assays from on-target activity screens. The data solely includes affinity data with pIC50 or pEC50 as endpoints. The assays were chosen to be representative, exhibiting various assay sizes and active ratios. Figure 1[A] summarizes the number of measurements and the ratio of actives for all assays used in our study. Standardized SMILES were obtained using the method described in the MELLODDY-TUNER [1] package and extended connectivity fingerprints (ECFPs) of size 1024 and radius 2 were generated with RDKit [8]. Given that the date of each measurement is available, a real temporal split was performed. After ordering the data according to the measurement date, the data was split into five folds of equal size, so that each fold represented a specific period in the assay history. For generating single-task classification models, two label types were used to assess if the incorporation of aleatoric uncertainty improves model performance. First, hard labels were generated using a pIC50/pEC50 threshold of 6 for assigning active or inactive labels based on the result. This specific threshold was chosen because the models will be deployed in the early stages of the drug discovery pipeline, in which the desired binding affinity of drug candidates is in the micromolar range ($10^{-6}$ molar concentration) corresponding to a pIC50/pEC50 of 6. Second, the same threshold was applied and the assay-specific measurement error, corresponding to the standard deviation of the control compound measurements, was used to obtain probabilistic labels. In detail, a normal distribution $X \sim \mathcal{N}(\mu, \sigma^2)$ was generated, where $\mu$ corresponded to the chosen threshold and $\sigma^2$ to the standard deviation of the control compound of the respective assay. In this step, the control compound corresponded to the compound with the most measurements in the respective assay. Subsequently, the CDF of these assay-specific distributions was used to obtain the probabilistic label [9]. Figure 1[B] shows the standard deviation (Std) of the control compound as well as the available number of measurements to calculate the Std for every assay.

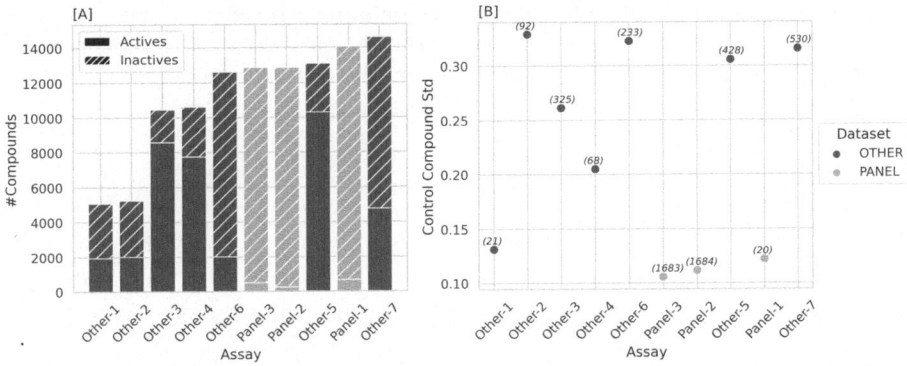

**Fig. 1. Overview over Assay data.** Assays from two categories, 'Other' and 'Panel', were used. [A] The number of measurements (#Compounds) of each assay as the sum of active and inactive compounds (pIC50/pEC50 threshold = 6) is shown. [B] shows the standard deviation (Std) and the number of measurements (in brackets) of the control compound for every assay.

## 2.1 Model Generation

Random Forest (RF) models were generated using scikit-learn. The maximum depth of the trees and the required number of estimators were tuned using a validation dataset. Probability-like outputs were generated by taking the ratio of decision trees in an RF that voted for a specific test instance to be active. Furthermore, Probabilistic Random Forests (PRF) [15] were generated using probabilistic labels as ground truth. A detailed description of the PRF training procedure can be found in Mervin et al. [9]. Post-hoc probability calibration techniques fit a calibration model to the raw scores of a classifier using a separate calibration dataset. In our work, we use the validation dataset for this step. Two uncertainty calibration approaches were used, namely Platt scaling [14] and Venn-ABERS (VA) predictors [18]. Platt scaling [14] involves fitting a logistic regression to the classification scores to counteract over- or underfitted uncertainty estimations. For calibration with VA predictors [18] two isotonic regression functions are trained on the validation data and the test instance, representing the two possible hypotheses that the test instance is active versus inactive. As such, two different probabilities are obtained from the isotonic regression models, corresponding to a lower and an upper bound on the probability, which are subsequently condensed to a point estimate as proposed by Tocatelli et al. [17]. For more detail on VA predictors we refer to Mervin et al. [10].

## 3  Results

### 3.1  Incorporation of Aleatoric Uncertainty Using Measurement Errors

Table 1 summarizes the Binary Cross Entropy (BCE↓) loss and the Adaptive Calibration Error (ACE↓) [12] for five model repeats of all model types trained on two example datasets, namely the Panel-1 and Other-3 assays. The first

**Table 1. Overview over RF model performance based on two example assays.** Averages over five model repeats are shown. The best results for each metric are marked in bold, while not significantly worse scores are indicated in italics.

| Method | Panel-1 | | Other-3 | |
|--------|---------|---------|---------|---------|
| | BCE ↓ | ACE ↓ | BCE ↓ | ACE ↓ |
| **Hard Labels** | | | | |
| RF | *0.187 ± 0.005* | 0.032 ± 0.001 | 0.312 ± 0.009 | 0.192 ± 0.008 |
| RF-Platt | **0.182 ± 0.004** | **0.029 ± 0.002** | 0.235 ± 0.007 | 0.117 ± 0.006 |
| RF-VA | *0.183 ± 0.002* | 0.037 ± 0.001 | **0.211 ± 0.007** | **0.089 ± 0.006** |
| **Probabilistic Labels** | | | | |
| PRF | *0.181 ± 0.002* | *0.032 ± 0.002* | 0.307 ± 0.01 | 0.187 ± 0.008 |
| PRF-Platt | **0.181 ± 0.001** | **0.03 ± 0.001** | 0.229 ± 0.005 | 0.112 ± 0.004 |
| PRF-VA | 0.185 ± 0.002 | 0.038 ± 0.002 | **0.212 ± 0.006** | **0.092 ± 0.006** |

three folds were used for model training, while the last fold was used for testing. Using probabilistic labels instead of hard labels improves the calibration error and the BCE loss of the RF and RF-Platt models trained on Other-3 assay data. Models for the Panel-1 assay do not show any improvements when incorporating aleatoric error. This result could be explained by the difference in standard deviations shown in Fig. 1[B], which are used for generating the probabilistic labels. Given that the measurement error of the Panel-1 assay is smaller compared to the Other-3 assay the normal distribution used for generating the probabilistic labels is narrower, resulting in probabilistic labels that are more similar to the hard labels, thus leading to similar results of RF and PRF models. The post-hoc calibration methods improve the BCE loss and ACE scores of Other-3 models, with RF-VA performing best in terms of both metrics, with a BCE and ACE of $0.211 \pm 0.007$ and $0.089 \pm 0.006$, respectively. The results for the Panel-1 assay show that in terms of ACE the RF-Platt model performs slightly better than the uncalibrated RF model, while the PRF models did not improve after calibration. In general, the control compounds of the Panel assays exhibit smaller standard deviations than those of the Other assays, as illustrated in Fig. 1[B]. The results of the assays omitted from Table 1 reveal that using probabilistic labels generally leads to better BCE scores for Other assays. In contrast, such clear improvements can not be observed for Panel assays. This could be a result of the differences mentioned above in standard deviations of the control compounds between the assay categories. However, there are also exceptions from this trend, such as the Other-1 assay, which does not show improvements when including probabilistic labels, despite the large standard deviation of its control compound. Hence, we conclude that it is required to look at the model performance on the individual assay to find the best calibration method for that specific dataset. For all assays, the same model performs best in terms of BCE scores when comparing models trained with hard labels versus probabilistic labels. This is also true in terms of the ACE results, except for the Other-1 assay, for which the RF model performs best for hard labels and the VA-calibrated model is best for probabilistic labels. However, the difference between PRF and PRF-VA is not significant. A more elaborate study is required to understand the effect of probabilistic labels on probability calibration in detail, which will be the object of our future research but is outside the scope of this abstract.

## 3.2   Probability Calibration Across Evolving Test Sets

Figure 2 shows the performance of five model repeats of different RF models across all ten assays and test sets in terms of ACE. The models were trained on one fold and then used for separately predicting three test folds representing subsequent time spans in the assay history. Test set 1 corresponds to the fold closest in time to the training fold, while test set 3 represents the fold furthest away. The ACE for test set 1 is the smallest across all models for the majority of assays as shown in Fig. 2, indicating that the models are better calibrated for compounds measured closer in time to the training fold. This pattern can also be observed in some assays when comparing test sets 2 and 3, however, the tendency is not as clearly visible as for test set 1. One of the reasons for the observed behavior could be a distribution shift in training and test data that increases as we progress in time, which is supported by a paper by Ovadia et al. [13], in which an increasing distribution shift was reported to impair probability calibration.

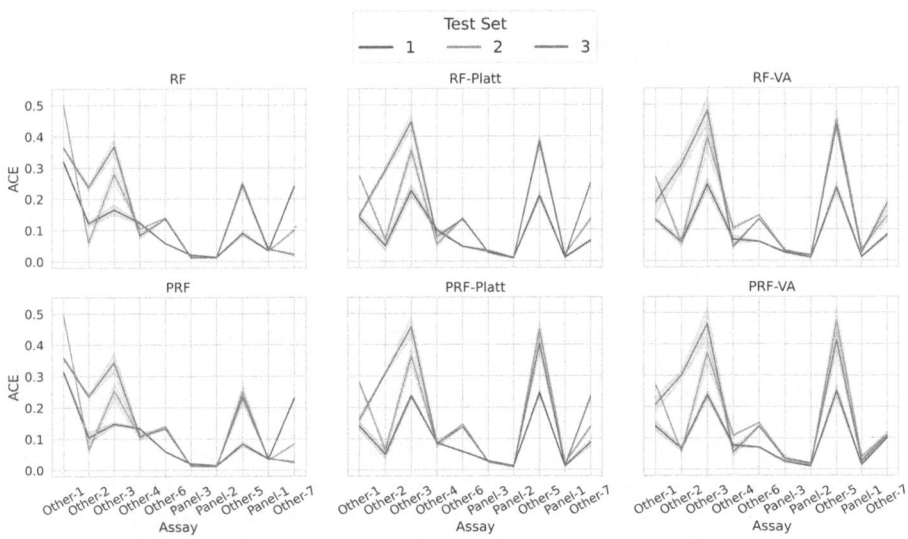

**Fig. 2. Model calibration over time.** The Adaptive Calibration Error (ACE) is shown for five model repeats across all assays. The models were trained on one training fold. Test Set 1 is closest in time to the training set, whereas Test Set 3 is furthest away.

## 4   Conclusion and Outlook

In this study, we showed that using probabilistic labels in combination with probability calibration approaches can improve uncertainty estimation in RF models. In addition, we present a comprehensive analysis of how model calibration changes over time using temporal splits of internal data from a pharmaceutical company. Based on these preliminary results, we will take further steps to understand model calibration in a temporal setting. Furthermore, we will extend our study to other model architectures, including multi-layer perceptrons (MLP), to investigate if the same conclusions can be drawn for other model types. Finally, we will explore uncertainty estimation methods to account for model uncertainty, including deep ensembles [7] and Monte-Carlo Dropout [2], to analyze if these approaches improve probability calibration.

**Acknowledgements.** Many thanks to Susanne Winiwarter of AstraZeneca in Gothenburg for her valuable advice during the data preparation. This study was partially funded by the European Union's Horizon 2020 research and innovation programme under the Marie Skłodowska-Curie Actions grant agreement "Advanced machine learning for Innovative Drug Discovery (AIDD)" No. 956832.

**Disclosure of Interests.** The authors have no competing interests to declare that are relevant to the content of this article.

## References

1. MELLODDY-TUNER. https://github.com/melloddy/MELLODDY-TUNER
2. Gal, Y., Ghahramani, Z.: Dropout as a bayesian approximation: representing model uncertainty in deep learning. In: Balcan, M.F., Weinberger, K.Q. (eds.) Proceedings of The 33rd International Conference on Machine Learning. Proceedings of Machine Learning Research, vol. 48, pp. 1050–1059. PMLR, New York, New York, USA (20–22 Jun 2016). https://proceedings.mlr.press/v48/gal16.html
3. Gruber, C., Schenk, P.O., Schierholz, M., Kreuter, F., Kauermann, G.: Sources of uncertainty in machine learning – a statisticians' view (2023). https://doi.org/10.48550/arXiv.2305.16703
4. Guo, C., Pleiss, G., Sun, Y., Weinberger, K.Q.: On calibration of modern neural networks. In: Precup, D., Teh, Y.W. (eds.) Proceedings of the 34th International Conference on Machine Learning. Proceedings of Machine Learning Research, 06–11 Aug 2017, vol. 70, pp. 1321–1330. PMLR (2017). https://proceedings.mlr.press/v70/guo17a.html
5. Heyndrickx, W., et al.: Melloddy: Cross-pharma federated learning at unprecedented scale unlocks benefits in qsar without compromising proprietary information. J. Chem. Inf. Model. (2023). https://doi.org/10.1021/acs.jcim.3c00799
6. Hüllermeier, E., Waegeman, W.: Aleatoric and epistemic uncertainty in machine learning: an introduction to concepts and methods. Mach. Learn. **110**, 457–506 (2019). https://doi.org/10.1007/s10994-021-05946-3

7. Lakshminarayanan, B., Pritzel, A., Blundell, C.: Simple and scalable predictive uncertainty estimation using deep ensembles. In: Guyon, I., et al. (eds.) Advances in Neural Information Processing Systems. vol. 30. Curran Associates, Inc. (2017), https://proceedings.neurips.cc/paper_files/paper/2017/file/9ef2ed4b7fd2c810847ffa5fa85bce38-Paper.pdf

8. Landrum, G.: RDKit: Open-source cheminformatics (2006). https://doi.org/10.5281/zenodo.6961488

9. Mervin, L., Trapotsi, M.A., Afzal, A., Barrett, I., Bender, A., Engkvist, O.: Probabilistic random forest improves bioactivity predictions close to the classification threshold by taking into account experimental uncertainty (2021). https://doi.org/10.26434/chemrxiv.14544291

10. Mervin, L.H., Afzal, A.M., Engkvist, O., Bender, A.: Comparison of scaling methods to obtain calibrated probabilities of activity for protein-ligand predictions. J. Chem. Inf. Model. **60**(10), 4546–4559 (2020). https://doi.org/10.1021/acs.jcim.0c00476, pMID: 32865408

11. Mervin, L.H., Johansson, S., Semenova, E., Giblin, K.A., Engkvist, O.: Uncertainty quantification in drug design. Drug Discovery Today **26**(2), 474–489 (2021). https://doi.org/10.1016/j.drudis.2020.11.027

12. Nixon, J., Dusenberry, M.W., Zhang, L., Jerfel, G., Tran, D.: Measuring calibration in deep learning. In: Proceedings of the IEEE/CVF Conference on Computer Vision and Pattern Recognition (CVPR) Workshops (June 2019). https://doi.org/10.48550/arXiv.1904.01685

13. Ovadia, Y., et al.: Can you trust your model' s uncertainty? Evaluating predictive uncertainty under dataset shift. In: Wallach, H., Larochelle, H., Beygelzimer, A., d' Alché-Buc, F., Fox, E., Garnett, R. (eds.) Advances in Neural Information Processing Systems, vol. 32. Curran Associates, Inc. (2019). https://proceedings.neurips.cc/paper_files/paper/2019/file/8558cb408c1d76621371888657d2eb1d-Paper.pdf

14. Platt, J.: Probabilistic outputs for support vector machines and comparisons to regularized likelihood methods. Adv. Large Margin Classif. **10**(3), 61–74 (1999)

15. Reis, I., Baron, D., Shahaf, S.: Probabilistic random forest: a machine learning algorithm for noisy data sets. Astron. J. **157**(1), 16 (2018). https://doi.org/10.3847/1538-3881/aaf101

16. Sheridan, R.P.: Time-split cross-validation as a method for estimating the goodness of prospective prediction. J. Chem. Inf. Model. **53**(4), 783–790 (2013). https://doi.org/10.1021/ci400084k

17. Toccaceli, P., Nouretdinov, I., Luo, Z., Vovk, V., Carlsson, L., Gammerman, A.: Excape wp1-probabilistic prediction (2016)

18. Vovk, V., Petej, I.: Venn-abers predictors (2014). https://doi.org/10.48550/arXiv.1211.0025

**Open Access** This chapter is licensed under the terms of the Creative Commons Attribution 4.0 International License (http://creativecommons.org/licenses/by/4.0/), which permits use, sharing, adaptation, distribution and reproduction in any medium or format, as long as you give appropriate credit to the original author(s) and the source, provide a link to the Creative Commons license and indicate if changes were made.

The images or other third party material in this chapter are included in the chapter's Creative Commons license, unless indicated otherwise in a credit line to the material. If material is not included in the chapter's Creative Commons license and your intended use is not permitted by statutory regulation or exceeds the permitted use, you will need to obtain permission directly from the copyright holder.

# Curating Reagents in Chemical Reaction Data with an Interactive Reagent Space Map

Mikhail Andronov[1,2]($\boxtimes$) ⓘ, Natalia Andronova[5], Michael Wand[1,3], Jürgen Schmidhuber[1,4] ⓘ, and Djork-Arné Clevert[2] ⓘ

[1] IDSIA, USI, SUPSI, 6900 Lugano, Switzerland
{mikhail.andronov,michael.wand}@idsia.ch
[2] Machine Learning Research, Pfizer Research and Development, Friedrichstr. 110, 10117 Berlin, Germany
djork-arne.clevert@pfizer.com
[3] Institute for Digital Technologies for Personalized Healthcare, SUPSI, 6900 Lugano, Switzerland
[4] AI Initiative, KAUST, 23955 Thuwal, Saudi Arabia
juergen.schmidhuber@kaust.edu.sa
[5] Independent Researcher, Berlin, Germany

**Abstract.** The increasing use of machine learning and artificial intelligence in chemical reaction studies demands high-quality reaction data, necessitating specialized tools enabling data understanding and curation. Our work introduces a novel methodology for reaction data examination centered on reagents - essential molecules in reactions that do not contribute atoms to products. We propose an intuitive tool for creating interactive reagent space maps using distributed vector representations, akin to word2vec in Natural Language Processing, capturing the statistics of reagent usage within datasets. Our approach enables swift assessment of reagent action patterns and identification of erroneous reagent entries, which we demonstrate using the USPTO dataset. Our contributions include an open-source web application for visual reagent pattern analysis and a table cataloging around six hundred of the most frequent reagents in USPTO annotated with detailed roles. Our method aims to support organic chemists and cheminformatics experts in reaction data curation routine.

**Keywords:** Reagents · word2vec · USPTO · Chemical data curation

## 1 Introduction

Over the many years chemical science has existed, chemists have amassed a vast body of knowledge and records about organic chemical reactions. This wealth of information encapsulates hundreds of distinct reaction types determined by a general transformation scheme and the required reagents [13]. As machine

© The Author(s) 2025
D.-A. Clevert et al. (Eds.): AIDD 2024, LNCS 14894, pp. 21–35, 2025.
https://doi.org/10.1007/978-3-031-72381-0_3

learning is cementing its place among widely used approaches to various reaction modeling problems [8,16], chemists are becoming increasingly concerned with understanding and curating their reaction data.

The most widely known reaction data collections are Reaxys [3], CASREACT [1], the open dataset of reactions from US patents (USPTO) [15], Pistachio [2], and a recent Open Reaction Database (ORD) [10]. Different reaction datasets may have unique particularities and biases [24], and comprehending those in one's dataset of choice is an advisable prerequisite for any study relying on that dataset. One of the possible sources of reaction data imperfections is reagent information.

A chemical reaction scheme (Fig. 1) typically involves reactants, products, and reagents. A reactant, as defined by IUPAC, is a substance consumed in the course of a chemical reaction. Consequently, reagents are other molecules that enable a reaction but do not contribute atoms to the products. Reagents are commonly written above or below the arrow in a reaction scheme. For example, catalysts and solvents are reagents. However, in practice, for convenience, substances with other roles, such as reducing and oxidizing agents, may also be considered reagents. Reagents may be integral to the mechanism of a reaction or merely improve the reaction rate. Reactions that use the same reagents may often correspond to the same reaction type.

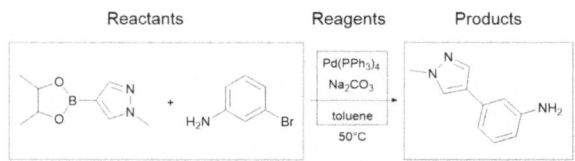

**Fig. 1.** An instance of the Suzuki coupling reaction. Reagents are normally written above or below the arrow. Various reaction types are often enabled by specific reagents.

Before building and testing machine learning models on some reaction dataset, it may be beneficial to pay special attention to reagent information within reagent records: verify whether reactant-reagent separation is adequate, or if there are redundant records of the same reagent, or if information about detailed roles of reagents is available. While the creators of popular reaction databases try to address all those issues, the data often needs additional curation. For example, reactant-reagent separation in USPTO is based on atom-atom mapping provided by the Indigo toolkit. The imperfection of this mapping may lead to imperfect separation. Additionally, reagent roles may not be detailed enough: both Reaxys and USPTO feature only three roles ("catalysts", "solvents", or "reagents"). When building reagent prediction models, their detailed performance analysis may require access to richer reagent role attribution [4].

In this paper, we describe a simple visual tool that helps to curate reagents in chemical reaction data. This tool is an interactive reagent space map based on distributed vector representations of reagents and served in a web application.

We obtain reagent representations using an algorithm equivalent to the famous word2vec [18] algorithm from Natural Language Processing, and they reflect the statistics of reagent co-occurrences and interaction in a given reaction data corpus: representations of reagents of the same role tend to cluster together. We demonstrate the application of our tool to the USPTO dataset [15]. Using our tool, we label around six hundred most common reagents in USPTO into ten detailed roles, detect reactants erroneously listed as reagents, and ensure the uniqueness of unique reagents' names.

Our reagent space mapping and the web application work with any reaction dataset, and we are confident that it will benefit organic chemists and cheminformatics specialists working with their own reaction data.

The codebase is open-source and available at https://github.com/Academich/reagent_emb_vis.

## 2    Results

### 2.1    Interactive Application

We have built a lightweight web application for the interactive exploration of the USPTO reagent space. Figure 2 demonstrates the application's appearance. The application displays an interactive UMAP [17] projection of reagent embeddings that capture the statistics of reagent co-occurrences within the dataset. Various filters are available in the application: it is possible to display only reagents with desired indices or roles, and one can also filter reagents by SMARTS patterns. When the user hovers over a point on the map, the corresponding reagent structure and its SMILES appear on the screen. Two versions of the map are available: a flat 2D map and a map on the surface of a sphere.

Users can explore reagents in their own reaction data in the application after carrying out necessary preprocessing described in the repository with the code.

### 2.2    Properties of Reagent Embeddings

If some two reagents tend to occur in similar contexts, i.e., with similar other reagents, then it is likely that these two reagents are alternatives, and they likely get embeddings that are close to each other. For example, we do not expect two different palladium catalysts for Suzuki coupling to occur in one reaction, but one can use the same bases and solvents with both of them. Therefore, we can expect the map of reagent embeddings to feature role clusters, e.g., a cluster of specific catalysts, ligands, or other reagents. The reagent embeddings are obtained by factorizing the point-wise mutual information (PMI) scores matrix with singular value decomposition (SVD). We can easily derive the table of PMI scores from the table of reagent counts by Eq. 1.

$$\mathrm{PMI}(x, y) = \log_2 \frac{P(x, y)}{P(x)P(y)} \tag{1}$$

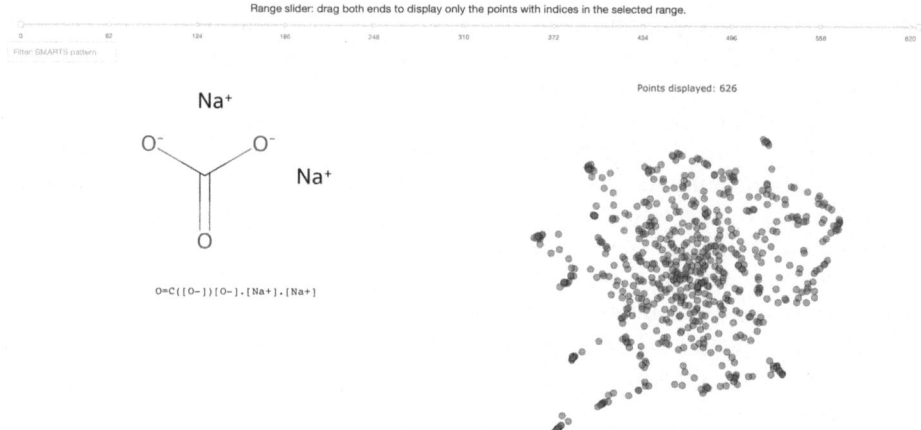

**Fig. 2.** The appearance of the web application for the exploration of reagent space. Reagents in our demonstration come from USPTO. On the right, there is an interactive UMAP projection of reagent embeddings. Every point corresponds to a unique reagent. Upon hovering over a point, the corresponding reagent is displayed on the left. Various filters for the map are available. The embeddings are obtained by factorizing the matrix of point-wise mutual information scores between pairs of reagents with singular value decomposition (SVD). The original embedding dimensionality is 50.

where $x$ and $y$ are reagents, $P(x,y)$ is a relative frequency of a reagent pair among all reagent pairs, and $P(x)$ and $P(y)$ are relative frequencies of individual reagents. Reagent vector representations obtained in this manner lie close in the vector space for entities with similar "meanings", which are determined by the "companions" of those entities. This method of obtaining reagent embeddings is equivalent to the word2vec algorithm (see 3.1) if we treat reagent molecules as words and reagents in one reaction as one context. One can freely select reagent embedding dimensionality, and we choose it to be 50 to achieve information compression that forces reagent embeddings into role clusters but with enough degrees of freedom. Other dimensionalities lead to different shapes of the map but do not tend to affect the observed clusters.

## 2.3    Reagent Data Curation

When reagents are displayed in this interactive map (Fig. 2), it is much easier to curate them than when working with just table data. We showcase two use cases when the map facilitates reagent curation: labeling reagents into detailed roles and finding redundancies in reagent SMILES entries.

**Reagent Number.** To study the reagents in USPTO, we obtain the entire dataset using the rxnutils [9] Python package and carry out relevant preprocessing (see 3.2). We rely on the atom-atom mapping (AAM) provided in USPTO to extract reagents from reactions. After filtering, we are left with 1,128,297 reactions that feature at least one reagent. In these reactions, we recognize 40,556 unique molecules as reagents using AAM. Among them, two-thirds (27,100) occur only once. We disregard them, as their attribution to reagents is mostly the result of erroneous atom-atom mapping, and, in any case, our method works best with reagents that occur in the data several times and desirably come together with various other reagent species. Furthermore, for demonstration we decided to limit our study to the reagents that occur at least 100 times in the filtered dataset, and we are left with 626 unique reagent SMILES, which we sort by occurrence frequency in descending order.

**Reagent Roles.** Reagent role information in the USPTO is rather limited and only differentiates between catalysts, solvents, and everything else. We decided to manually categorize every reagent in our subset of 626 USPTO reagents into one of the following eleven roles:

**Acids**
Acidic compounds typically used as catalysts, e.g., HCl, $H_2SO_4$.

**Bases**
Basic compounds typically used as catalysts, e.g., NaOH, n-butyllithium, or Hünig's base.

**Lewis Acids**
Catalysts that are Lewis acids, e.g., $AlCl_3$.

**Catalysts**
Other catalysts, mostly metal-based. For example, those would comprise homogeneous palladium-based catalysts for cross-coupling reactions, such as Suzuki coupling, or heterogeneous catalysts for hydrogenation, such as metal nickel.

**Ligands**
Compounds with the purpose of forming coordination complexes with metal ions, e.g., phosphorus-based ligands for homogeneous palladium catalysts or chelating agents.

**Oxidizing Agents**
Various oxidizers, including halogenating agents. For example, $KMnO_4$, $CrO_3$, $SOCl_2$.

**Reducing Agents**
Various reducing agents, e.g. $H_2$, $SnCl_2$, $LiAlH_4$.

**Activators**
Reagents that facilitate an overall reaction but are consumed in the process. For example, we call activators the agents that allow the formation of active intermediates enabling a reaction, such as active esters for peptide coupling reactions. A comprehensive review by El-Faham and Albericio [5] systematizes a number of such reagents, many of which are present in USPTO. The examples of activators are 1-hydroxybenzotriazole and N,N'-dicyclohexylcarbodiimide for peptide coupling or diethyl azodicarboxylate for Mitsunobu reaction.

**Ambience**
Other reagents that do not fit any described role and serve some auxiliary purpose, such as radical reaction inhibitors or nitrogen for inert atmosphere.

**Reactants**
Molecules that, in fact, contribute atoms to the product and are mistakenly classified as reagents.

The interactivity of the application and the clustering tendency of reagents with the same role on the map allow for faster decision-making about reagent roles compared to using only tabular reagent data. After reagent labeling, the map in the web application looks as in Fig. 3. The reagents are now colored according to the roles we assigned to them.

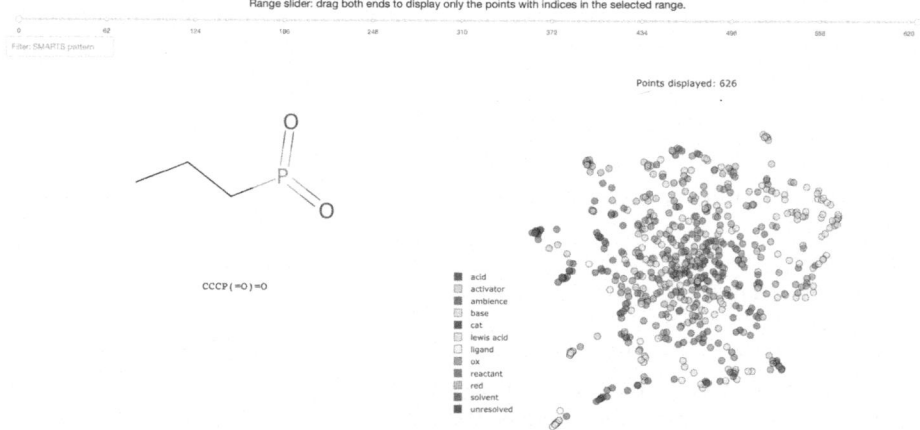

**Fig. 3.** The map of embeddings for the subset of 626 most common USPTO reagents colored according to the detailed reagent roles assigned manually.

The tendency to role cluster formation is visible in the map, although it is not perfect.

**Questionable Reagent SMILES.** The properties of the reagent embedding map enable easy detection of different SMILES that represent the same reagent. Alternative SMILES of the same reagent are naturally assigned similar embeddings by our method. For example, Fig. 4 demonstrates a zoom-in on a map region occupied by strong bases.

By exploring this region on the interactive map, we can immediately discover that there are sometimes several different SMILES representations for the same reagent in USPTO. We see two SMILES representations for n-butyllithium, two for lithium diisopropylamide, and three for lithium bis(trimethylsilyl)amide. In other regions of the map in Fig. 3, there are sometimes mixtures of reagents determined in our preprocessing as one reagent (see 3.2), e.g., two solvents together

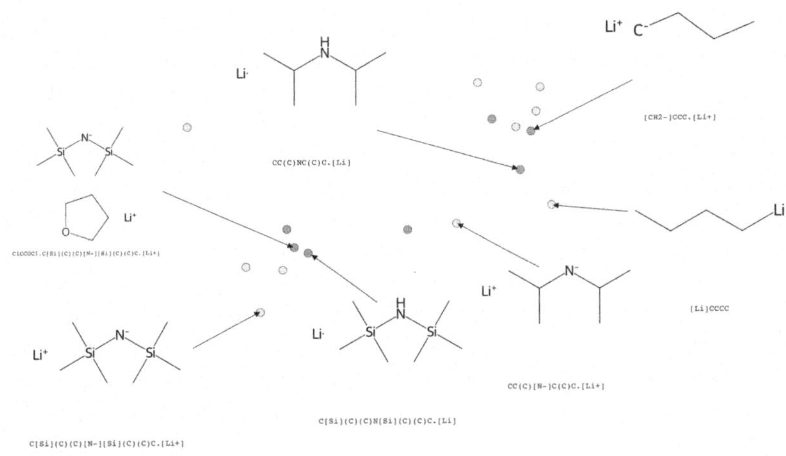

**Fig. 4.** A region of the reagent embedding map that reveals unique reagents represented by several different SMILES. In this region of strong bases, the SMILES representations of n-butyllithium, lithium diisopropylamide, and lithium bis(trimethylsilyl)amide require standardization.

as a unique reagent. We want to ensure that each unique reagent is represented by a unique SMILES string in our reaction data. We revise our reagent map once more, standardizing all redundant SMILES we find this way and reducing the number of unique reagent entries in our map from 626 to 559.

## 2.4   Analysis

After labeling our reagents with roles and standardizing all reagent SMILES, we generate reagent embeddings again. The embedding map we thus obtain has a slightly different shape but preserves the same clusters united by reagent action.

**Contiguous Regions of Reagent Action.** Figure 5 demonstrates the Voronoi diagram for the reagent embedding map we obtain for the final 559 reagents in our application. A Voronoi diagram for a set of two-dimensional points is a partition of a plane into regions drawn around every point in the set. In this case, the points are called seeds, and the regions are called Voronoi cells. In every cell, the points of the plane are closer to the seed forming the cell than to any other seed. In our case, UMAP projections of reagent embeddings are seeds.

The cells formed by reagent embedding projections are colored by the roles of the corresponding reagents, and the touching cells of the same color merge. A Voronoi diagram makes it easy to see regions formed by reagents of the same purpose. We highlight nine example regions in the diagram. The region labeled with 1 comprises various reagents enabling peptide coupling. Among them are HOBt, HOAt, DCC, and their alternatives, many of which are described in the corresponding review [5]. Region 2 unites organophosphorus ligands for homogeneous metal catalysts, e.g., 1,3-bis(diphenylphosphino)propane (dppp), CyJohnPhos and BrettPhos. Region 3 is defined by homogeneous palladium catalysts, such as $Pd2(dba)_3$ or $Pd(PPh_3)_4$. Region 4 is the region of chelating agents, e.g., 8-hydroxyquinoline or phenanthroline. Region 5 features two clusters: one cluster is defined by hydrogenation catalysts, such as palladium on carbon and other 10th group metals; the other cluster is formed by catalytic compounds of $Cu^I$ and $Cu^{II}$. Region 6 unites the reagents for Mitsunobu reactions, namely diethyl azodicarboxylate (DEAD) and structurally similar TMAD, DIAD, and DtBAD. Region 7 comprises chlorinating agents such as $SOCl_2$, $PCl_5$, or cyanuric chloride. Region 8 is the region of Grignard reagents. Region 9 features borohydrides that serve as reducing agents, e.g., $NaBH_4$ or $NaBH(Ac)_3$. These nine regions are just some examples, and the map contains many more regions uniting reagents of the same action. We invite the readers to explore the interactive map of USPTO reagents themselves.

**Reagent Role Distribution.** Figure 6 demonstrates the distribution of reagent roles among our subset of 559 reagents.

One can see that nineteen percent of the reagents are actually reactants. It hints that the atom mapping tool used in USPTO often fails to resolve noisy reactions this dataset is notorious for.

We make the list of USPTO reagents and their roles available alongside our codebase. We hope that researchers will this information useful for their own work involving USPTO.

**Reagent Counts.** Figure 7 displays the logarithm of the number of occurrences for every reagent. Around 50 most common reagents dominate the dataset. However, $\sim$ 35 percent of all reactions use reagents other than those 50 most common ones. The less common reagents form a fat tail of the occurrence frequency distribution: the relative occurrence frequency of reagents starting from the 100th falls like $\frac{1}{x^2}$.

## 3    Methods

### 3.1    Theory

The original word2vec [18] is a machine learning algorithm, whose success popularized machine learning for Natural Language Processing. The goal of the algorithm is to obtain learned distributed vector representations of words of the natural language for subsequent usage in downstream tasks, such as text classification [14]. It has been also shown that embeddings similar to word2vec enable very effective text compression [21]. The idea behind word2vec is the distributional hypothesis: the words occurring in similar contexts have similar meanings and, therefore, must get similar vector representations - embeddings. The algorithm iteratively trains those initially random embeddings by solving a classification task - what words are in the context for a given word. The context

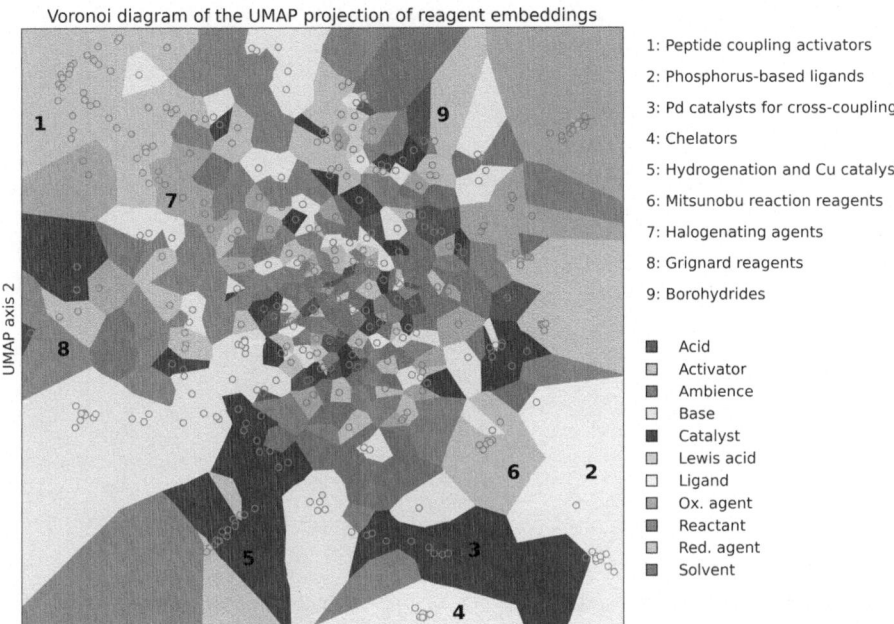

**Fig. 5.** Voronoi diagram of the UMAP projection of reagent embeddings. Reagents of the same role tend to form contiguous regions corresponding to the same type of reagent action. Numbers highlight 9 example regions that unite reagents of the same purpose.

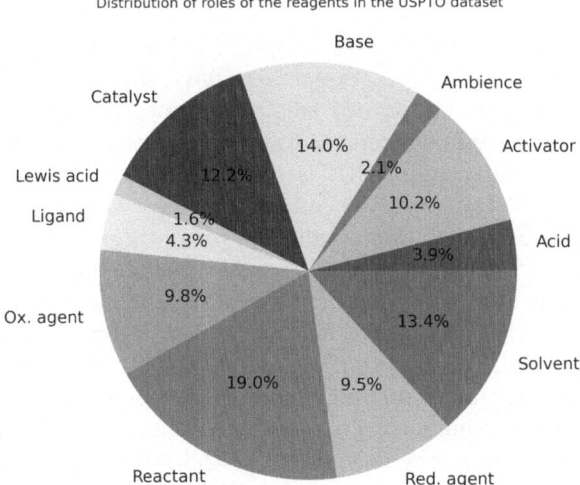

**Fig. 6.** Distribution of 559 most common USPTO reagents by detailed roles.

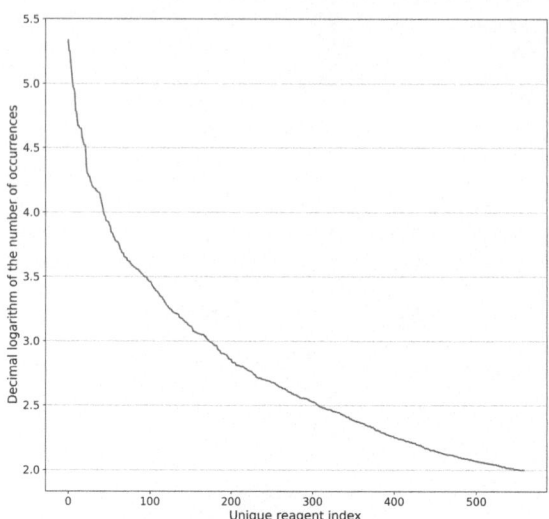

Occurrence distribution (truncated to 100) for every unique reagent in the USPTO dataset

**Fig. 7.** Decimal logarithm of the number of occurrences of unique reagents in the USPTO dataset. We consider only the reagents that occur at least 100 times. The dataset is dominated by approximately 50 most common reagents, and the others are relatively rare. Unique reagent indices are sorted by occurrence frequency.

of a word consists of other words within a window centered on that word, and the data pairs for training are obtained using a sliding window over a large text corpus.

Word2vec has gained widespread adoption in the fields of cheminformatics and bioinformatics. For example, researchers have adapted it for the construction of universal feature vectors for small molecules [7,23]. Also, it has been employed to create meaningful representations of nucleic acids for phylogenetic analysis [20], predicting drug-miRNA associations [6] and RNA degradation prediction [11]. Additionally, word2vec embeddings have been utilized for proteins in tasks such as drug-target interaction [26], drug-target affinity [28], protein-protein interaction [27], and others. A survey [19] offers a broad overview of the diverse applications of word2vec in bioinformatics and cheminformatics.

While word2vec is an iterative algorithm, Levy and Golberg [12] derived a proof that a formulation of word2vec called SGNS (skipgram with negative sampling) is equivalent to factorising the matrix of point-wise mutual information (PMI) scores with singular value decomposition (SVD). PMI scores can be obtained from co-occurrence counts (Eq. 1). Although learning word embeddings is better approached with the iterative algorithm as there are hundreds of thousands of unique words and their co-occurrence matrix would be too large, we can resort to PMI matrix factorization when building reagent embeddings, as there are only several hundreds of the latter.

Singular value decomposition is a way of factorising a matrix in linear algebra. It is a generalization of matrix eigendecomposition to non-square matrices. SVD decomposes a real-valued matrix $M$ of size $m \times n$ into three factors according to Eq. 2:

$$M = U \Sigma V^T \tag{2}$$

where $U$ and $V$ are orthogonal matrices with sizes $m \times m$ and $n \times n$, respectively, and $\Sigma$ is an $m \times n$ diagonal matrix with non-negative real numbers on the diagonal called singular values. The number of non-zero singular values is equal to the rank of $M$, and we can truncate the sizes of $\Sigma$ to the desired number of singular values, which controls the sizes of $U$ and $V$. We perform SVD in Python using the SciPy library [25] and use $U$ as the matrix of reagent embeddings, in which every row is a reagent embedding with the dimensionalily equal to the chosen number of singular values.

## 3.2   Data

**Preprocessing.** We use the entire USPTO dataset for our study. We download the USPTO reactions as SMILES using the rxnutils Python package [9]. The raw data before preprocessing consists of 3,748,191 reactions. We use rxnutils to build the pipeline for the initial preprocessing. First, we canonicalize all SMILES strings and remove the reactions that contain molecules not canonicalizeable by RDKit. Then, we remove stereochemical information from all molecules and drop duplicate reactions. Finally, we keep only those reactions in which the number of precursors (reactants and reagents together) is less than 10. This initial preprocessing reduces the data volume to 1,393,677 reactions. We do not remove the CXSMILES information because it is necessary to assemble scattered reagent fragments like ions or ligands that belong together into one species.

The consequent preprocessing does not rely on rxnutils and consists of the following steps:

1. Reagent extraction:
   We determine reagents for every reaction using the atom-atom mapping information available in USPTO. We use the CXSMILES information provided in reactions to assemble disjoint reagent fragments (e.g., ions or ligands) into whole reagent species.
2. Reaction filtering:
   We drop reaction records involving more than ten reactants and reagents, more than five reagents, or zero reagents. We also delete trivial reactions - the ones in which the product is found among reactants or reagents.
3. Reagent filtering:
   We remove bound water from reagents such as $Na_2SO_4 \cdot 10H_2O$ and delete bare ions and other reagent species with unbalanced charge.

Eventually, we remove reactants and products from every reaction and obtain a text file in which every row contains reagent SMILES for a reaction separated by semicolons. In this file, we count all unique reagents and remove those that appear less than 100 times. This leaves us with 626 unique reagents (559 after ensuring one-to-one correspondence between unique reagents and unique SMILES in our data, see 2.3).

We then use this file to derive a table of reagent co-occurrence counts and point-wise mutual information scores. Finally, we factorize this table using singular value decomposition and use one of the factors as a matrix of reagent embeddings.

**UMAP Details.** We use the default parameters of the UMAP class (15 neighbors, Euclidean metric) in the UMAP Python package [17] in the web application.

**Alternative Reagent Determination.** As an alternative to relying on AAM to determine reagents, we also try the fingerprint-based procedure described by Schneider et al. [22] and available in RDKit. It does not depend on AAM and is therefore more universal, even though it may occasionally fail, determining all reactants as reagents. In such cases, we fall back to the AAM-based reagent extraction. With this preprocessing, we obtain 558 unique reagents that appear at least 100 times in the dataset. Among those, 25 reagents, 14 of which we assign the "reactant" role, are not among our 559 reagents determined using AAM. At the same time, 26 reagents (19 "reactants") determined by AAM are not in the set of reagents determined by the fingerprint procedure. However, the reagent embedding space maps in both cases do not differ significantly, and the map for reagents obtained by the fingerprint procedure contains the same regions as in Fig. 5. Therefore, we conclude that both reagent determination procedures are alternatives for the user to choose from depending on the user's confidence in the AAM reliability in their data.

## 4  Conclusion

Our paper introduces a novel approach to facilitate chemical reaction data curation with a focus on reagents. By counting unique reagents in a reaction dataset, turning the table of their pairwise counts into point-wise mutual information scores, and factorizing that table with singular value decomposition, we effectively apply a word2vec algorithm to reagents and obtain their distributed vector representations that capture reagent co-occurrence statistics. Projecting the obtained reagent representations on the plane with UMAP, we construct a reagent space map demonstrating intriguing clustering patterns among reagents, highlighting that reagents united by common purpose lie close together and partition into distinct clusters. Based on this map, we present an interactive web application providing a user-friendly platform for researchers to navigate and explore reagent patterns within reaction datasets. We demonstrate the use of the application with the USPTO dataset.

Additionally, we systematize and catalog several hundreds of the most common reagents used in USPTO and label them into detailed roles. We believe that such information will be valuable for reagent prediction models trained on USPTO. For example, it can be used to estimate the performance of a reagent prediction model not by the often interchangeable individual molecules, but by the correctness of predicted roles. The code and data are available at https:// github.com/Academich/reagent_emb_vis.

**Acknowledgments.** This study was partially funded by the European Union's Horizon 2020 research and innovation program under the Marie Skłodowska-Curie Innovative Training Network European Industrial Doctorate grant agreement No. 956832 "Advanced machine learning for Innovative Drug Discovery, and also by the Horizon Europe funding programme under the Marie Skłodowska-Curie Actions Doctoral Networks grant agreement "Explainable AI for Molecules - AiChemist" No. 101120466.

**Disclosure of Interests.** The authors have no competing interests to declare.

# References

1. CASREACT website. Accessed 23 March 2024. https://www.cas.org/cas-data
2. NextMove Software. Pistachio. Accessed 23 March 2023. http://www.nextmovesoftware.com/pistachio.html
3. Reaxys database. Accessed 23 March 2024. https://www.reaxys.com
4. Andronov, M., Voinarovska, V., Andronova, N., Wand, M., Clevert, D.A., Schmidhuber, J.: Reagent prediction with a molecular transformer improves reaction data quality. Chem. Sci. **14**(12), 3235–3246 (2023)
5. El-Faham, A., Albericio, F.: Peptide coupling reagents, more than a letter soup. Chem. Rev. **111**(11), 6557–6602 (2011)
6. Guan, Y.J., et al.: MFIDMA: a multiple information integration model for the prediction of drug-miRNA associations. Biology **12**(1), 41 (2022)
7. Jaeger, S., Fulle, S., Turk, S.: Mol2vec: unsupervised machine learning approach with chemical intuition. J. Chem. Inf. Model. **58**(1), 27–35 (2018)
8. Johansson, S., et al.: AI-assisted synthesis prediction. Drug Discov. Today Technol. **32**, 65–72 (2019)
9. Kannas, C., Genheden, S.: Rxnutils–a cheminformatics python library for manipulating chemical reaction data (2022)
10. Kearnes, S.M., et al.: The open reaction database. J. Am. Chem. Soc. **143**(45), 18820–18826 (2021)
11. Krishna, U.V., Premjith, B., Soman, K.: A comparative study of pre-trained gene embeddings for COVID-19 mRNA vaccine degradation prediction. In: Proceedings of the Seventh International Conference on Mathematics and Computing: ICMC 2021, pp. 301–308. Springer (2022)
12. Levy, O., Goldberg, Y.: Neural word embedding as implicit matrix factorization. Adv. Neural Inf. Proc. Syst. **27** (2014)
13. Li, J.J.: Name Reactions, 3rd edn. A Collection of Detailed Reaction Mechanisms. Springer-Verlag, Berlin Heidelberg (2006). https://doi.org/10.1007/978-3-030-50865-4
14. Lilleberg, J., Zhu, Y., Zhang, Y.: Support vector machines and word2vec for text classification with semantic features. In: 2015 IEEE 14th International Conference on Cognitive Informatics & Cognitive Computing (ICCI* CC), pp. 136–140. IEEE (2015)

15. Lowe, D.M.: Extraction of Chemical Structures and Reactions from the Literature. Ph.D. Dissertation, University of Cambridge, Cambridge, UK. https://doi.org/10. 17863/CAM.16293 (2012)
16. Madzhidov, T.I., et al.: Machine learning modelling of chemical reaction characteristics: yesterday, today, tomorrow. Mendeleev Commun. **31**(6), 769–780 (2021)
17. McInnes, L., Healy, J., Melville, J.: UMAP: uniform manifold approximation and projection for dimension reduction. arXiv preprint arXiv:1802.03426 (2018)
18. Mikolov, T., Sutskever, I., Chen, K., Corrado, G.S., Dean, J.: Distributed representations of words and phrases and their compositionality. Adv. Neural Inf. Proc. Syst. **26** (2013)
19. Öztürk, H., Özgür, A., Schwaller, P., Laino, T., Ozkirimli, E.: Exploring chemical space using natural language processing methodologies for drug discovery. Drug Discov. Today **25**(4), 689–705 (2020)
20. Ren, R., Yin, C., S.-T. Yau, S.: kmer2vec: A novel method for comparing DNA sequences by word2vec embedding. J. Comput. Biol. **29**(9), 1001–1021 (2022)
21. Schmidhuber, J., Heil, S.: Sequential neural text compression. IEEE Trans. Neural Networks **7**(1), 142–146 (1996)
22. Schneider, N., Stiefl, N., Landrum, G.A.: What's what: the (nearly) definitive guide to reaction role assignment. J. Chem. Inf. Model. **56**(12), 2336–2346 (2016). https://doi.org/10.1021/acs.jcim.6b00564
23. Shao, J., Gong, Q., Yin, Z., Pan, W., Pandiyan, S., Wang, L.: S2DV: converting SMILES to a drug vector for predicting the activity of anti-HBV small molecules. Brief. Bioinform. **23**(2) (2022)
24. Thakkar, A., Kogej, T., Reymond, J.L., Engkvist, O., Bjerrum, E.J.: Datasets and their influence on the development of computer assisted synthesis planning tools in the pharmaceutical domain. Chem. Sci. **11**(1), 154–168 (2020)
25. Virtanen, P., et al.: SciPy 1.0 Contributors: SciPy 1.0: Fundamental Algorithms for Scientific Computing in Python. Nat Methods **17**, 261–272 (2020). https://doi. org/10.1038/s41592-019-0686-2
26. Wang, L., Zhou, Y., Chen, Q.: AMMVF-DTI: a novel model predicting drug-target interactions based on attention mechanism and multi-view fusion. Int. J. Mol. Sci. **24**(18), 14142 (2023)
27. Wang, Y., You, Z.H., Yang, S., Li, X., Jiang, T.H., Zhou, X.: A high efficient biological language model for predicting protein-protein interactions. Cells **8**(2), 122 (2019)
28. Xia, M., Hu, J., Zhang, X., Lin, X.: Drug-target binding affinity prediction based on graph neural networks and word2vec. In: International Conference on Intelligent Computing, pp. 496–506. Springer (2022). https://doi.org/10.1007/978-3-031-13829-4_43

**Open Access** This chapter is licensed under the terms of the Creative Commons Attribution 4.0 International License (http://creativecommons.org/licenses/by/4.0/), which permits use, sharing, adaptation, distribution and reproduction in any medium or format, as long as you give appropriate credit to the original author(s) and the source, provide a link to the Creative Commons license and indicate if changes were made.

The images or other third party material in this chapter are included in the chapter's Creative Commons license, unless indicated otherwise in a credit line to the material. If material is not included in the chapter's Creative Commons license and your intended use is not permitted by statutory regulation or exceeds the permitted use, you will need to obtain permission directly from the copyright holder.

# Latent-Conditioned Equivariant Diffusion for Structure-Based De Novo Ligand Generation

Julian Cremer[1,2(✉)], Tuan Le[1,3], Djork-Arné Clevert[1], and Kristof T. Schütt[1]

[1] Pfizer Research and Development, Friedrichstraße 110, 10117 Berlin, Germany
[2] Universitat Pompeu Fabra, Biomedical Research Park (PRBB), C Dr. Aguiader 88, 08003 Barcelona, Spain
jn.cremer@icloud.com
[3] Freie Universität Berlin, Arnimallee 12, 14195 Berlin, Germany

**Abstract.** We propose PoLiGenX for *de novo* ligand design using latent-conditioned, target-aware equivariant diffusion. Our model leverages the conditioning of the generation process on reference molecules within a protein pocket to produce shape-similar *de novo* ligands that can be used for target-aware hit expansion and hit optimization. The results of our study showcase the efficacy of PoLiGenX in ligand design. Docking scores indicate that the generated ligands exhibit superior binding affinity compared to the reference molecule while preserving the shape. At the same time, our model maintains chemical diversity, ensuring the exploration of diverse chemical space. The evaluation of Lipinski's rule of five suggests that the sampled molecules possess a higher drug-likeness than the reference data. This constitutes an important step towards the controlled generation of therapeutically relevant *de novo* ligands tailored to specific protein targets.

**Keywords:** Equivariant Graph Neural Networks · Diffusion · Generative Chemistry · Structure-based drug discovery · De novo molecule design · Hit Expansion

## 1 Introduction

In recent years, the intersection of artificial intelligence (AI) and drug discovery has witnessed remarkable strides, with the potential to revolutionize the traditional approaches to identifying novel therapeutic compounds. Among these innovations, AI-enabled structure-based drug discovery has emerged as a promising research avenue, in particular in form of equivariant target-aware diffusion models. By conditioning the diffusion process on the receptors of proteins, these models exhibit a remarkable capacity to generate *de novo* ligands with enhanced

---

J. Cremer and T. Le—Contributed equally to this work.

© The Author(s) 2025

D.-A. Clevert et al. (Eds.): AIDD 2024, LNCS 14894, pp. 36–46, 2025.
https://doi.org/10.1007/978-3-031-72381-0_4

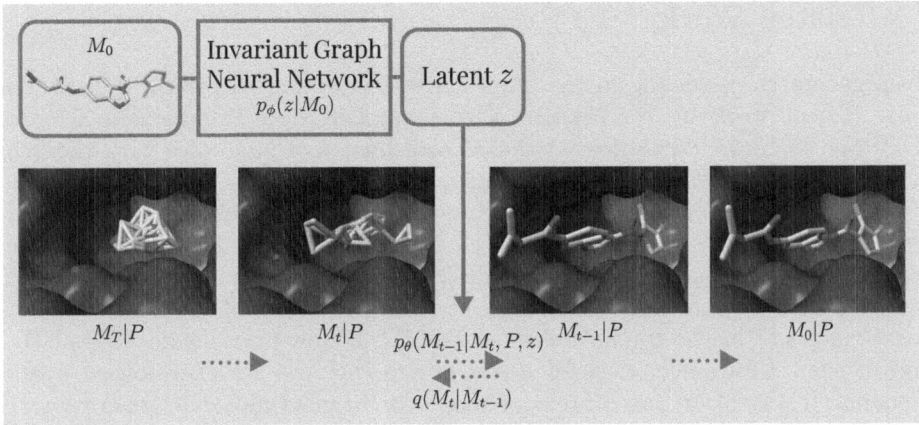

**Fig. 1.** Graphical model of the proposed latent diffusion model. The encoded ligand $z$ serves as input to the diffusion model $p_\theta$ to steer the generation process of new ligands $M_0$.

affinity [7, 14, 19, 20]. Failing to consider the essential chemical properties for target binding can lead to a significant lack of specificity and result in ineffective drug candidates. Moreover, these candidates must exhibit favorable absorption, distribution, metabolism, excretion (ADME), and toxicity profiles. Designing ligands from scratch without addressing these critical properties may produce molecules with poor bioavailability or potential toxicity, thereby limiting their therapeutic potential. This challenge is further exacerbated by the often sparse and noisy data available for developing effective machine learning models. However, machine learning shows considerable promise during the hit expansion phase of drug discovery. This crucial stage involves enhancing and exploring the chemical space around promising hits already identified through high-throughput screening or other methods. In this study, we introduce PoLiGenX (**Po**cket-based **Li**gand **Gen**erator for hit e**X**pansion) that generates ligands *de novo* within a protein binding pocket. Unlike previous models, PoLiGenX starts with a seed molecule, such as a hit candidate or an initial scaffold, and iteratively refines and modifies it to improve its efficacy. We enhance the capabilities of the existing equivariant diffusion model, EQGAT-diff [14], by incorporating a latent encoding as a condition. It is derived from an invariant graph neural network that is jointly trained to process 3D molecular inputs. The setup ensures that the newly generated ligands retain structural characteristics of the seed molecules while undergoing necessary chemical modifications and diversification. Our proposed approach adds a new level of control to the process of generating *de novo* ligands, aligning it more closely with the specific needs of targeted drug design, particularly during the hit expansion phase (Fig. 1).

## 2    Related Work

Deep generative modeling in the life sciences has become a promising research area. Recent work by [12,25] uses Denoising Diffusion Probabilistic Models (DDPMs) [8,13,21,22] to predict the 3d coordinates of molecules with the help of 3d equivariant graph neural networks. In the *de novo* setting, another line of research focuses on directly generating the atomic coordinates and elements, using autoregressive models [5,6,18]. [9] introduced E(3) equivariant diffusion model (EDM) for *de novo* molecule design that simultaneously learns atomic elements next to the coordinates while treating chemical elements as continuous variables to utilize the formalism of DDPM. Follow-up work leverages EDM and develop diffusion models for linker design [11] and structure-based ligand modeling [7,14,20]. In the context of shape-conditioned molecule generation [1] (SQUID) and [3] (ShapeMol) recently proposed to incorporate the shape of a seed molecule into the generation process. Both approaches use an equivariant surface encoding of a seed molecule, whereby SQUID uses variational auto-encoding on graphs and focuses more on fragment-based design. ShapeMol is an adaption of SQUID in 3d space leveraging an equivariant diffusion model. However, both works do not include a protein receptor condition. We propose to use a simple approach employing reference molecules in a latent representation, as outlined in more details below.

## 3    Methods

*Problem Formulation and Notation.* We investigate the generation of molecular structures $M$ in a *de novo* setting conditioned on a protein pocket $P$, i.e., building a generative model $p_\theta(M|P)$. For this, we use the EQGAT-diff framework proposed by [14]. In this setup, a noisy ligand $M_t = (X_t, H_t, E_t)$ - representing perturbed atomic coordinates, element types, and bond features - is used, and the diffusion model $p_\theta$ predicts the uncorrupted data modalities $(\hat{X}_0, \hat{H}_0, \hat{E}_0)$, because the distribution $M_{t-1}|M_t$ depends on both $M_t$ and $\hat{M}_0$. Specifically, for continuous coordinates, the reverse distribution adheres to a multivariate Gaussian model, while for discrete-valued modalities, it follows a categorical distribution. We refer to [14] for further details.

While models like EQGAT-diff, TargetDiff or DiffSBDD generate ligands in context of a protein pocket, they do not constraint the generated ligands to preserve properties like shape or chemical similarity during training. In contrast, we include a latent variable $z \in \mathbb{R}^K$ that relates to the input molecule $\hat{M}_0$. The latent $z$ may serve as a shape conditioning that also comprises chemical information like the atom composition of the molecule $\hat{M}_0$. The graph encoder $q_\phi : \mathcal{X}^M \to \mathbb{R}^K$ is invariant to permutation, rotation and translation of atoms [15,24].

Following [1], chemical similarity of two molecules is measured as the Tanimoto similarity of ECFP4 fingerprints (2048 bits) computed by RDKit, whereby shape similarity is defined by Gaussian descriptions of molecular shape in form of atom-centered Gaussians and calculated by the volume overlaps between them as in [1].

*PoLiGenX.* To model the dependence on variable $z$, we include a variational distribution $q_\phi(z|M_0)$ similar to [17, 26] and obtain the ELBO

$$p_\theta(M_0|P) = \mathbb{E}_{q(M_{1:T}|M_0)q_\phi(z|M_0)} \left[ \frac{p_\theta(M_0, M_{1:T}, z|P)}{q(M_{1:T}|M_0)q_\phi(z|M_0)} \right]$$

$$\geq \mathbb{E}_{q(M_{1:T}|M_0)q_\phi(z|M_0)} \left[ \log \frac{p_\theta(M_0, M_{1:T}, z|P)}{q(M_{1:T}|M_0)q_\phi(z|M_0)} \right]$$

$$= \mathbb{E}_{q(M_1|M_0)q_\phi(z|M_0)} [\log p_\theta(M_0|M_1, P, z)] \qquad (1)$$

$$+ \mathbb{E}_{q(M_T|M_0)q_\phi(z|M_0)} \left[ \log \frac{p(M_T|z)}{q(M_T|M_0)} \right]$$

$$- D_{KL}(q_\phi(z|M_0)||p(z)) - \sum_{t=2}^{T} \mathbb{E}_{q(M_t|M_0)q_\phi(z|M_0)} [L_{t-1}],$$

where the diffusion loss $L_{t-1}$ is per timestep and defined as

$$L_{t-1} = D_{KL}(q(M_{t-1}|M_t, M_0)||p_\theta(M_{t-1}|M_t, P, z)). \qquad (2)$$

We extend the diffusion model by a conditioning on $z$ and train $p_\theta(M|P, z)$ to minimize the KL divergence to the tractable reverse distribution, which is achieved when predicting the original data points $\hat{M}_0$ [2,8,14]. Similar to prior works, we optimize the diffusion $L_{t-1}$ by drawing steps per minibatch instead of the entire trajectory. We adopt a Gaussian prior for the latent distribution, i.e., $p(z) \sim N(0, I)$ and enforce a smooth latent space by choosing the maximum mean discrepancy (MMD) loss [23] over the KL divergence. The prior distribution for the ambient data space, i.e., $M_T$ is a 0-CoM Gaussian for coordinates and empirical categorical distribution for discrete data types from the training set as discussed in [14]. During training, we sample a batch of pocket-ligand pairs and a step $t \in \{1, \ldots, 500\}$. Next, we encode the ligands $M_0$ into latents $z$, apply the noise process to the ligands to obtain $M_t$ and minimize the diffusion loss while providing $z$ as an additional input via adaptive layer normalization [10] next to the protein pocket $P$. We refer to the supplementary material for further details including the derivation of the ELBO.

## 4   Results

We train PoLiGenX using the CrossDocked2020 [4] dataset, following the same dataset splits as found in previous research [7,14,16,19,20]. Unlike other models, PoLiGenX incorporates not only the protein pocket as a condition for generating novel ligands but also utilizes a latent embedding of a ligand from the dataset as an initial condition. This distinctive approach positions PoLiGenX differently from the mentioned models - it is specifically designed to perform tasks akin to

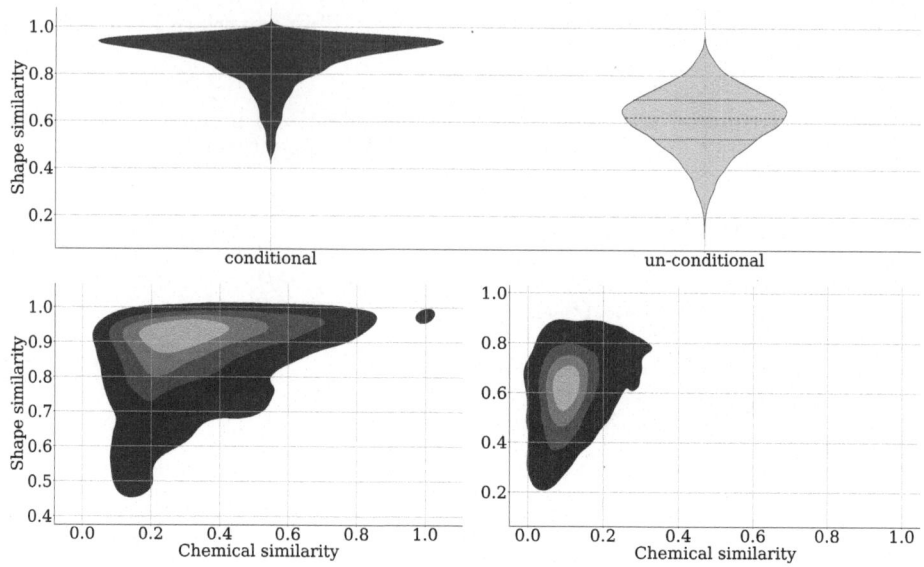

**Fig. 2. Top**: Violin plot of the Tanimoto shape similarity evaluated across all test targets of the CrossDocked dataset. PoLiGenX (left) is compared to EQGAT-diff (right). In the conditional setting the model generates significantly more shape-similar molecules. **Bottom**: Heatmap histogram comparing PoLiGenX (left) with EQGAT-diff with respect to Tanimoto shape and chemical similarity on the CrossDocked test set. The brighter the color the higher the molecule count.

hit expansion by enhancing specificity, chemical diversity, and binding affinity, rather than operating solely as a target-aware, but unconditional *de novo* model. In the following, we evaluate if PoLiGenX effectively maintains the structural shape of the seed molecule while promoting chemical diversity.

Figure 2 (top) shows the evaluation of the mean shape similarity on the Cross-Docked test set for both PoLiGenX (conditional) and EQGAT-diff (unconditional). The test set comprises 100 ligand-pocket complexes for which 100 ligands each were sampled and the Tanimoto shape similarity measured against the reference ligands. PoLiGenX exhibits significantly higher shape similarities across complexes. However, we aim to preserve the shape between reference and sample *without* sacrificing chemical diversity to ensure an efficient exploration of chemical space. Figure 2 (bottom) shows the distribution of shape similarity against chemical similarity for conditional and unconditional sampling. We observe a mean shape similarity of 0.64 and 0.12 chemical similarity for EQGAT-diff. In contrast, PoLiGenX exhibits a significant increase in shape similarity with mean value of 0.87, but also generates a reasonably high diversity in samples with mean chemical similarity of 0.33.

To evaluate the expressiveness of the learned latent embeddings, Fig. 3 visualizes the UMAP projections of the latent embeddings. We sampled 100 ligands per receptor for ten randomly selected targets of the CrossDocked test set. The

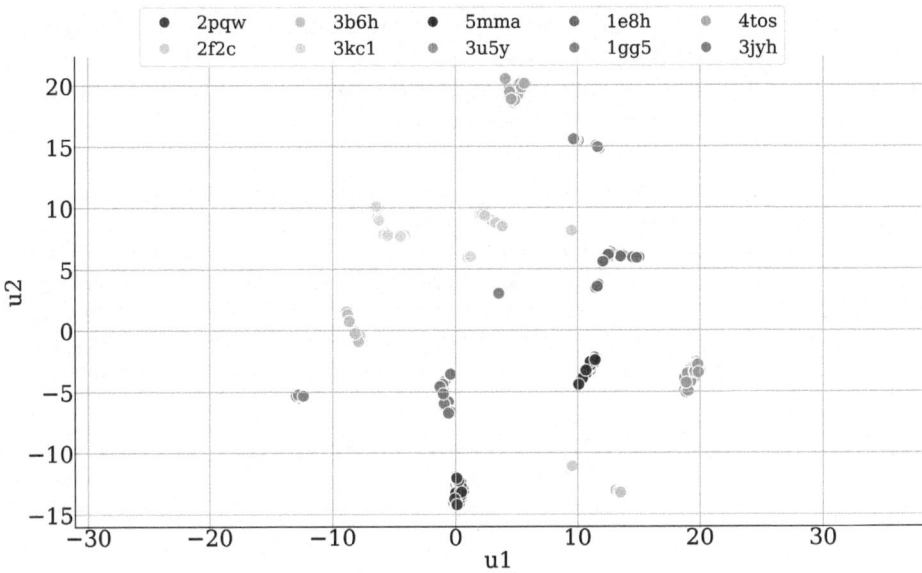

**Fig. 3.** UMAP plot showing the 2d projections of the latent embeddings of 100 sampled ligands per target for ten randomly sampled test set targets.

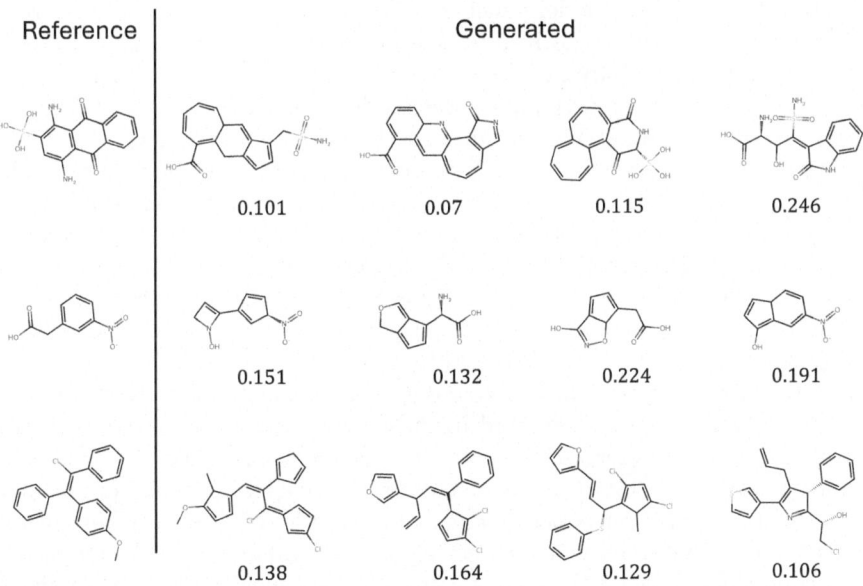

**Fig. 4.** Reference molecules extracted from the CrossDocked test split (left) and four generated molecules sampled randomly with PoLiGenX. Below each generated ligand, we also show the chemical similarity to the reference ligand.

resulting UMAP projections reveal that the latent embeddings effectively separate the ligands into distinct clusters specific to each target. This observation suggests that our latent model successfully captures the context of ligands in relation to their respective protein receptors.

**Table 1.** Docking performance on the CrossDocked test set and ligands generated using PoLiGenX. QuickVina2 is employed for docking. We report mean values across all targets with standard deviations given as subscripts. Drug-likeness is measured via RDKit's QED value. Further, molecules are evaluated in terms of the octanol-water partition coefficient (logP), the molecular weight (MolWt) and the number of hydrogen acceptors and donors. Following Lipinski's rule of five, we report the percentage of molecules that obey the respective rule. The last column gives the average of molecules fulfilling all rules.

| Data | QVina2 (All) ↓ | QVina2 (Top-10%) ↓ | QED ↑ | logP ↑ | MolWt ↑ | H-acceptors ↑ | H-donors ↑ | Lipinski ↑ |
|---|---|---|---|---|---|---|---|---|
| CrossDocked test set | $-6.85_{\pm 2.33}$ | – | $0.47_{\pm 0.20}$ | 0.79 | 0.85 | 0.84 | 0.8 | $3.35_{\pm 1.14}$ |
| PoLiGenX | $\mathbf{-7.21}_{\pm 2.22}$ | $\mathbf{-8.04}_{\pm 2.44}$ | $\mathbf{0.59}_{\pm 0.20}$ | **0.91** | **0.87** | **0.85** | **0.91** | $\mathbf{3.57}_{\pm 0.93}$ |

Next, we compare molecules sampled conditionally from our model, PoLiGenX, with the reference test data, focusing on docking scores and chemical properties. As previously outlined, the purpose of PoLiGenX is significantly different to recent *de novo* models, such as EQGAT-diff, hence we omit a comparison. Table 1 summarizes the results. We observe improved docking scores for generated samples compared to the CrossDocked test data, in particular within the top 10% of each target. Here, we reach a docking score of $-8.04 \pm 2.44$ compared to $-6.85 \pm 2.33$ for the test data. At the same time, the generated ligands per target show improvement in RDKit's drug-likeness score (QED) and adherence to Lipinski's Rule of Five. These are chemical features recognized from a medicinal chemistry perspective as guidelines to identify compounds likely to possess favorable bioavailability. Specifically, the octanol-water partition coefficient (logP) should be less than 5, molecular weight (MolWt) should be less than 500 Daltons, hydrogen bond acceptors (H-acceptors) less than 10 and hydrogen bond donors (H-donors) should be less than 5.

Figure 4 depicts three randomly chosen test set ligands with four conditionally sampled and randomly selected ligands each. Judging by visual inspection, the topology is well preserved. We note that chemical similarity, especially based on fingerprints can change drastically if some chemical elements are interchanged. As shown in the bottom panel in Fig. 2, PoLiGenX achieves a mean chemical similarity of around 0.33 while preserving shape similarity of 0.87 compared to the unconditional case with 0.12 and 0.64 for chemical and shape similarity, respectively.

The controllable generation of PoLiGenX can be further regulated by including a control parameter $\lambda \in (0, 1]$ that scales the latent $z$ when going into the diffusion model. That is, for small $\lambda$ values approaching 0, PoLiGenX does not include any latent information and collapses to the unconditional EQGAT-diff and only leverages the pocket information as context. With $\lambda$ interpolating

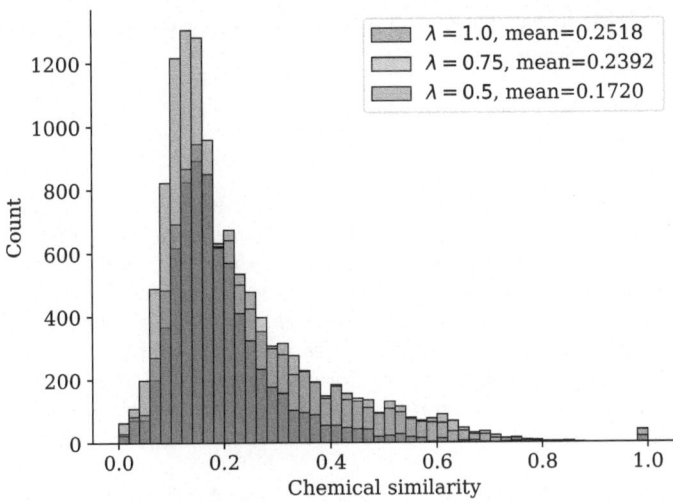

**Fig. 5.** Density plot for chemical similarity of generated ligands from PoLiGenX with varying $\lambda$ control parameter. With increasing $\lambda$, the latent $z$ of reference/seed ligand $M_0$ is preserved such that generated ligands exhibit higher chemical similarity to $M_0$.

between e.g. $(0.5, 1.0)$, we observe that the mean chemical similarity for generated ligands with respect to the references also increases as depicted in Fig. 5. We detail the influence of the latent variable $z$ in combination with the scale parameter $\lambda$ in the supplementary materials.

## 5   Conclusions

We have developed PoLiGenX for controlled *de novo* ligand generation within a protein binding pocket. By incorporating a latent encoding from a seed molecule into the diffusion model, we ensure that the generated ligands preserve shape and also adhere to the structural constraints of the target protein binding site. The effectiveness of PoLiGenX is evidenced by improved docking scores compared to reference ligands. Additionally, the generated ligands conform to Lipinski's Rule of Five, demonstrating their drug-likeness. Importantly, the model maintains chemical diversity, which is essential for exploring a broad range of chemical space and discovering novel therapeutic candidates. This integration of shape preservation, target specificity, and chemical diversity provides a powerful approach for the targeted generation of drug candidates, particularly useful in the hit expansion phase of drug discovery campaigns.

**Acknowledgments.** This study was partially funded by the European Union's Horizon 2020 research and innovation program under the Marie Skłodowska-Curie Innovative Training Network European Industrial Doctorate grant agreement No. 956832 "Advanced machine learning for Innovative Drug Discovery."

**Disclosure of Interests.** The authors have no competing interests to declare that are relevant to the content of this article.

# References

1. Adams, K., Coley, C.W.: Equivariant shape-conditioned generation of 3D molecules for ligand-based drug design. In: The Eleventh International Conference on Learning Representations (2023). https://openreview.net/forum?id=4MbGnp4iPQ
2. Austin, J., Johnson, D.D., Ho, J., Tarlow, D., van den Berg, R.: Structured denoising diffusion models in discrete state-spaces. In: Beygelzimer, A., Dauphin, Y., Liang, P., Vaughan, J.W. (eds.) Advances in Neural Information Processing Systems (2021). https://openreview.net/forum?id=h7-XixPCAL
3. Chen, Z., Peng, B., Parthasarathy, S., Ning, X.: Shape-conditioned 3D molecule generation via equivariant diffusion models (2023). https://arxiv.org/abs/2308.11890
4. Francoeur, P.G., Masuda, T., Sunseri, J., Jia, A., Iovanisci, R.B., Snyder, I., Koes, D.R.: Three-dimensional convolutional neural networks and a cross-docked data set for structure-based drug design. J. Chem. Inf. Model. **60**(9), 4200–4215 (2020). https://doi.org/10.1021/acs.jcim.0c00411, https://doi.org/10.1021/acs.jcim.0c00411
5. Gebauer, N., Gastegger, M., Schütt, K.: Symmetry-adapted generation of 3D point sets for the targeted discovery of molecules. In: Wallach, H., Larochelle, H., Beygelzimer, A., d'Alché-Buc, F., Fox, E., Garnett, R. (eds.) Advances in Neural Information Processing Systems, vol. 32. Curran Associates, Inc. (2019). https://proceedings.neurips.cc/paper_files/paper/2019/file/a4d8e2a7e0d0c102339f97716d2fdfb6-Paper.pdf
6. Gebauer, N.W.A., Gastegger, M., Hessmann, S.S.P., Müller, K.R., Schütt, K.T.: Inverse design of 3D molecular structures with conditional generative neural networks. Nat. Commun. **13**(1), 973 (2022). https://doi.org/10.1038/s41467-022-28526-y
7. Guan, J., Qian, W.W., Peng, X., Su, Y., Peng, J., Ma, J.: 3D equivariant diffusion for target-aware molecule generation and affinity prediction. In: The Eleventh International Conference on Learning Representations (2023). https://openreview.net/forum?id=kJqXEPXMsE0
8. Ho, J., Jain, A., Abbeel, P.: Denoising diffusion probabilistic models. In: Larochelle, H., Ranzato, M., Hadsell, R., Balcan, M., Lin, H. (eds.) Advances in Neural Information Processing Systems. vol. 33, pp. 6840–6851. Curran Associates, Inc. (2020). https://proceedings.neurips.cc/paper_files/paper/2020/file/4c5bcfec8584af0d967f1ab10179ca4b-Paper.pdf
9. Hoogeboom, E., Satorras, V.G., Vignac, C., Welling, M.: Equivariant diffusion for molecule generation in 3D. In: Chaudhuri, K., Jegelka, S., Song, L., Szepesvari, C., Niu, G., Sabato, S. (eds.) Proceedings of the 39th International Conference on Machine Learning. Proceedings of Machine Learning Research, vol. 162, pp. 8867–8887. PMLR (17–23 Jul 2022). https://proceedings.mlr.press/v162/hoogeboom22a.html
10. Huang, X., Belongie, S.: Arbitrary style transfer in real-time with adaptive instance normalization. In: 2017 IEEE International Conference on Computer Vision (ICCV), pp. 1510–1519 (2017). https://doi.org/10.1109/ICCV.2017.167

11. Igashov, I., et al.: Equivariant 3D-conditional diffusion models for molecular linker design. ArXiv **abs/2210.05274** (2022). https://arxiv.org/abs/2210.05274

12. Jing, B., Corso, G., Chang, J., Barzilay, R., Jaakkola, T.S.: Torsional diffusion for molecular conformer generation. In: Oh, A.H., Agarwal, A., Belgrave, D., Cho, K. (eds.) Advances in Neural Information Processing Systems (2022). https://openreview.net/forum?id=w6fj2r62r_H

13. Kingma, D.P., Salimans, T., Poole, B., Ho, J.: On density estimation with diffusion models. In: Beygelzimer, A., Dauphin, Y., Liang, P., Vaughan, J.W. (eds.) Advances in Neural Information Processing Systems (2021). https://openreview.net/forum?id=2LdBqxc1Yv

14. Le, T., Cremer, J., Noé, F., Clevert, D.A., Schütt, K.: Navigating the design space of equivariant diffusion-based generative models for de novo 3D molecule generation. In: The Twelfth International Conference on Learning Representations (2024). https://openreview.net/forum?id=kzGuiRXZrQ

15. Le, T., Noe, F., Clevert, D.A.: Representation learning on biomolecular structures using equivariant graph attention. In: The First Learning on Graphs Conference (2022). https://openreview.net/forum?id=kv4xUo5Pu6

16. Luo, S., Guan, J., Ma, J., Peng, J.: A 3d generative model for structure-based drug design. In: Ranzato, M., Beygelzimer, A., Dauphin, Y., Liang, P., Vaughan, J.W. (eds.) Advances in Neural Information Processing Systems, vol. 34, pp. 6229–6239. Curran Associates, Inc. (2021). https://proceedings.neurips.cc/paper_files/paper/2021/file/314450613369e0ee72d0da7f6fee773c-Paper.pdf

17. Luo, S., Hu, W.: Diffusion probabilistic models for 3D point cloud generation. In: Proceedings of the IEEE/CVF Conference on Computer Vision and Pattern Recognition (CVPR), pp. 2837–2845 (June 2021). https://openaccess.thecvf.com/content/CVPR2021/html/Luo_Diffusion_Probabilistic_Models_for_3D_Point_Cloud_Generation_CVPR_2021_paper.html

18. Luo, Y., Ji, S.: An autoregressive flow model for 3D molecular geometry generation from scratch. In: International Conference on Learning Representations (2022). https://openreview.net/forum?id=C03Ajc-NS5W

19. Peng, X., Luo, S., Guan, J., Xie, Q., Peng, J., Ma, J.: Pocket2Mol: Efficient molecular sampling based on 3D protein pockets. In: Chaudhuri, K., Jegelka, S., Song, L., Szepesvari, C., Niu, G., Sabato, S. (eds.) Proceedings of the 39th International Conference on Machine Learning. Proceedings of Machine Learning Research, vol. 162, pp. 17644–17655. PMLR (17–23 Jul 2022). https://proceedings.mlr.press/v162/peng22b.html

20. Schneuing, A., et al.: Structure-based drug design with equivariant diffusion models (2023). https://arxiv.org/abs/2210.13695

21. Sohl-Dickstein, J., Weiss, E., Maheswaranathan, N., Ganguli, S.: Deep unsupervised learning using nonequilibrium thermodynamics. In: Bach, F., Blei, D. (eds.) Proceedings of the 32nd International Conference on Machine Learning. Proceedings of Machine Learning Research, vol. 37, pp. 2256–2265. PMLR, Lille, France (07–09 Jul 2015). https://proceedings.mlr.press/v37/sohl-dickstein15.html

22. Song, Y., Sohl-Dickstein, J., Kingma, D.P., Kumar, A., Ermon, S., Poole, B.: Score-based generative modeling through stochastic differential equations. In: International Conference on Learning Representations (2021). https://openreview.net/forum?id=PxTIG12RRHS

23. Tolstikhin, I., Bousquet, O., Gelly, S., Schoelkopf, B.: Wasserstein auto-encoders. In: International Conference on Learning Representations (2018). https://openreview.net/forum?id=HkL7n1-0b

24. Winter, R., Bertolini, M., Le, T., Noe, F., Clevert, D.A.: Unsupervised learning of group invariant and equivariant representations. In: Oh, A.H., Agarwal, A., Belgrave, D., Cho, K. (eds.) Advances in Neural Information Processing Systems (2022). https://openreview.net/forum?id=47lpv23LDPr
25. Xu, M., Yu, L., Song, Y., Shi, C., Ermon, S., Tang, J.: Geodiff: a geometric diffusion model for molecular conformation generation. In: International Conference on Learning Representations (2022). https://openreview.net/forum?id=PzcvxEMzvQC
26. Zeng, X., Vahdat, A., Williams, F., Gojcic, Z., Litany, O., Fidler, S., Kreis, K.: LION: latent point diffusion models for 3D shape generation. In: Oh, A.H., Agarwal, A., Belgrave, D., Cho, K. (eds.) Advances in Neural Information Processing Systems (2022). https://openreview.net/forum?id=tHK5ntjp-5K

**Open Access** This chapter is licensed under the terms of the Creative Commons Attribution 4.0 International License (http://creativecommons.org/licenses/by/4.0/), which permits use, sharing, adaptation, distribution and reproduction in any medium or format, as long as you give appropriate credit to the original author(s) and the source, provide a link to the Creative Commons license and indicate if changes were made.

The images or other third party material in this chapter are included in the chapter's Creative Commons license, unless indicated otherwise in a credit line to the material. If material is not included in the chapter's Creative Commons license and your intended use is not permitted by statutory regulation or exceeds the permitted use, you will need to obtain permission directly from the copyright holder.

# Leveraging Quantum Mechanical Properties to Predict Solvent Effects on Large Drug-Like Molecules

Mathias Hilfiker[1,2]($\boxtimes$) (iD), Leonardo Medrano Sandonas[3] (iD), Marco Klähn[2] (iD), Ola Engkvist[2] (iD), and Alexandre Tkatchenko[1] (iD)

[1] Department of Physics and Materials Science, University of Luxembourg, 1511 Luxembourg City, Luxembourg
[2] Molecular AI, Discovery Sciences, R&D, AstraZeneca, Gothenburg 43183, Sweden
mathias.hilfiker@astrazeneca.com
[3] Institute for Materials Science and Max Bergmann Center of Biomaterials, TU Dresden, 01062 Dresden, Germany
leonardo.medrano@tu-dresden.de

**Abstract.** Understanding how solvation affects structure-property and property-property relationships of drug-like molecules is crucial for *de novo* design, as most relevant reactions occur in aqueous environments. We have thus performed an exhaustive analysis of the recently proposed Aquamarine dataset to gain insights into the effect of solvent-molecule interaction on the quantum-mechanical (QM) properties of large drug-like molecules. Our results show that the inclusion of an implicit solvent model of water changes the values of (extensive and intensive) QM properties but it does not alter the correlations among them. Moreover, we have found that solvation can limit the identification of unique molecular conformations, with variations in specific properties being rationalized by the extent of structural changes. $\Delta$-learning approach was used to predict solvent effects on the dipole moment $\mu$ and the many-body dispersion energy $E_{\mathrm{MBD}}$, resulting in more accurate and scalable predictive models compared to these directly trained on solvated properties. Hence, our work provides valuable insights into the effect of solvent-molecule interaction on physicochemical properties, which could assist in the development of machine-learning models for designing solvated molecules of pharmaceutical and biological relevance.

**Keywords:** Drug-like molecules · Quantum-mechanical properties · Solvent effects · Property prediction

## 1 Introduction

Solvation constitutes a big challenge for current chemoinformatics methods. A proper computational description of solvated systems requires explicit treatment of numerous interactions of different natures (*e.g.*, electrostatic, van der Waals,

© The Author(s) 2025
D.-A. Clevert et al. (Eds.): AIDD 2024, LNCS 14894, pp. 47–57, 2025.
https://doi.org/10.1007/978-3-031-72381-0_5

hydrogen bonding), which is unfeasible for large systems and long time scales. At the same time, however, accounting for solvent effects is crucial in many applications, from drug discovery to material science, because the aforementioned interactions can radically affect the electronic and geometrical structure of the solute, leading to large modifications of molecular and atomic properties. To appreciate these effects on large systems, models for implicit solvation have been proposed [1,2]. In these models, the solvent is considered as a continuum medium, and interactions are treated in a mean-field way, thus leading to a substantial reduction in degrees of freedom. This increased computational efficiency allows for simulations of larger systems and longer time scales compared to explicit methods while maintaining an acceptable accuracy for many applications. Examples of implicit models include modified Poisson-Boltzmann (MPB) [3], conductor-like screening model for real solvents (COSMO-RS) [4], and Generalized Born (GB) [5] model augmented with the hydrophobic solvent accessible surface area (GBSA) [6].

Even though the molecular theory of solvation is well-known [7], an exhaustive analysis of the effects that interactions with a solvent may have on an extensive set of solutes is still lacking. This is due to the vastness and diversity of the chemical space, which makes it extremely difficult to acquire comprehensive knowledge on the subject. The literature is rich in studies about solvent effects on specific systems, for example, Matczak *et al.* [8] considered ten formaldehyde and thioformaldehyde derivatives and studied how geometry, energetics, HOMO and LUMO orbitals, dipole moment and polarizability change upon solvation on solvents of low polarity. Instead, Odey *et al.* [9] studied how polar solvation affects the structure, dynamical stability, spectroscopy, and antiviral inhibitory potential of Cissampeline. Ensing *et al.* [10] also studied solvation effects on the $S_N2$ reaction between $Cl^-$ and $CH_3Cl$. This non-exhaustive list demonstrates how previous studies have concentrated on specific systems, properties, or reactions.

Within the context discussed above, in this work, we use the recently proposed Aquamarine (AQM) dataset [11] to obtain valuable insights into the effect of solvent-molecule interactions on structure/property and property/property relationships of large drug-like molecules. AQM contains 40 physicochemical properties for 59,783 conformers of 1,653 drug-like molecules with sizes up to 92 atoms, including H,C,N,O,F,P,S, and Cl elements. Using the two versions of AQM dataset, namely AQM-gas and AQM-sol, we have performed an extensive data-driven analysis of how hydration affects correlations between properties and energy ranking of conformers. We also investigated the relationship between the variation in properties (dipole moment $\mu$ and many-body dispersion energy $E_{MBD}$) and the structural changes. As a final analysis, we show that a machine learning model trained to learn the difference between the values of properties in water $P_{sol}$ and the values of the same properties in gas-phase $P_{gas}$ (*i.e.*, $\Delta P = P_{sol} - P_{gas}$) can help predict solvated-phase properties with higher precision compared to a model directly trained on $P_{sol}$ values. Previous studies used $\Delta$-learning to improve the level of theory of calculated properties [12–15], or to predict solvation free energies [16–18]. To the best of our knowledge, this is the first time $\Delta$-learning is applied to predict the QM properties of solvated molecules using only information from the gas-phase system.

## 2 Computational Methods

**Aquamarine Property Space.** Compounds in AQM [11] were sampled from ChEMBL [19] with the requirement of being similar to the Johnson & Johnson Innovative Medicines corporate database in terms of molecular weights (less than 1200), number of rotatable bonds (less than 30), quantitative estimate of drug-likeness (QED) score [20] (more than 0.4), and number of heavy atoms (less than 200). More details on the selection of representative chemistries can be found in Ref. [11]. For the selected molecules, conformers were generated using the conformational search workflow implemented in CREST code [21]. Representative conformers were optimized using the DFTB3 [22–24] method augmented with many-body dispersion (MBD) interactions [25–28]. Over 40 global and local properties were then computed at the PBE0+MBD [25, 29, 30] level of theory. The great advancement of AQM is its availability of these properties in gas-and solvated-phase, producing the two subsets AQM-gas and AQM-sol, respectively. For AQM-sol, conformations were optimized in implicit water using the GBSA [6] model. The properties calculations were instead performed using the MPB [3, 31] model. Among all the properties in AQM, we selected the the molecular dipole moment $\mu$ (an intensive property) and the many-body dispersion energy $E_{\mathrm{MBD}}$ (an extensive property) for further analysis. This choice is motivated by the fact that $\mu$ is expected to be heavily influenced by a polar solvent such as water. Whereas, $E_{\mathrm{MBD}}$ was selected because van der Waals interactions significantly affect the conformations of large molecules, so large changes in dispersion energy are expected to be related to significant changes in the 3D configuration of the molecule. These properties describe the electrostatic and long-range dispersion interactions in large molecular complexes, which are crucial components for screening the stability of molecules within drug discovery frameworks.

**Delta Learning Model.** We have used the state-of-the-art equivariant neural network MACE [32] to train the predictive models on $\Delta P = P_{\mathrm{sol}} - P_{\mathrm{gas}}$, with $P = \mu$ and $E_{\mathrm{MBD}}$. MACE decomposes the target property into atomic contributions and learns them as a function of atomic numbers and coordinates. In the final layer, these contributions are summed up to reconstruct the predicted property value. The training features were atomic numbers and cartesian coordinates in the gas phase. With the same features, other MACE models were trained to learn directly $P_{sol}$ values. We have trained four different models depending on the maximum number of atoms of molecules in the training data, i.e., up to 30 atoms (6,433 conformations), 40 atoms (12,760 conformations), 50 atoms (23,221 conformations), and 70 atoms (51,958 atoms). The validation set included 20% of the training data, and the evaluation was always performed on molecules larger than 70 atoms (6,625 conformations). To train the models, we have considered a cutoff radius of 6 Å and 2 MACE layers.

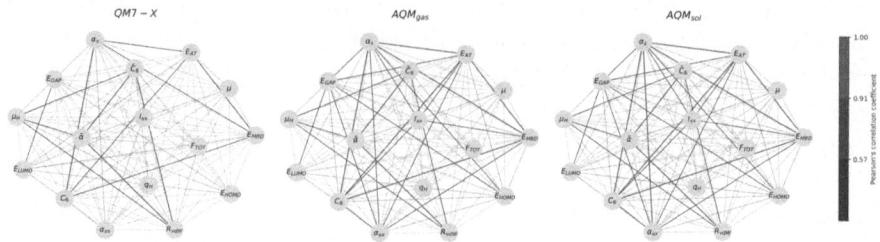

**Fig. 1.** Analysis of the correlation structure for small drug-like molecules in gas-phase (QM7-X), large drug-like molecules in gas-phase (AQM-gas) as well as in implicit water (AQM-sol). The thickness of edges is according to Pearson correlation coefficient $\rho$ (see color bar), and the color is purple for weakly correlated properties ($\rho < 0.57$), blue for moderately correlated properties ($0.57 \leq \rho < 0.91$), and green for strongly correlated properties ($\rho \geq 0.91$). (Color figure online)

## 3    Results and Discussion

We start by analyzing the pairwise property correlation among a set of selected QM properties contained in QM7-X [33], AQM-gas, and AQM-sol (see Fig. 1). As it was discussed by Medrano-Sandonas *et al.* [34], the lack of correlation among the majority of properties for small molecules contained in QM7-X is evidence of the "freedom of design" conjecture in the QM7-X property space, *i.e.*, there is the possibility of designing molecules with the desired value for a property, without sensibly affecting other properties [35]. To investigate the validity of this statement for large solvated drug-like molecules, we have performed a similar analysis using both AQM-gas and AQM-sol. Accordingly, we have examined pairwise correlations between 16 of out 19 properties studied in Ref. [34] (AQM dataset do not contain RMSD with respect to equilibrium structure, maximum distance between heavy atoms, and total DFTB energy). The selection criteria aimed to cover properties relevant to the drug discovery community from a quantum-mechanical point of view, striking a good balance between extensive and intensive, molecular and atomic, as well as ground-state and response properties. Following the approach in Ref. [34], we defined three categories according to $\rho$ values: weak correlations ($\rho < 0.57$), moderate correlations ($0.57 \leq \rho < 0.91$), and strong correlations ($\rho \geq 0.91$). As depicted in Fig. 1, we found out that "freedom of design" conjecture still exists (properties are generally weakly correlated), however, the number of moderately correlated pair of properties strongly increases from QM7-X to AQM (from 8 to 18). Interestingly, this result is only due to the increased molecular size, as the correlation structure observed in AQM-sol is largely similar to that in AQM-gas. This indicates that while solvation impacts the property values, it does not alter the correlations between them.

We proceeded with the analysis studying how solvation affects the energetic landscape of a molecule. To do this, we first computed the energy range for

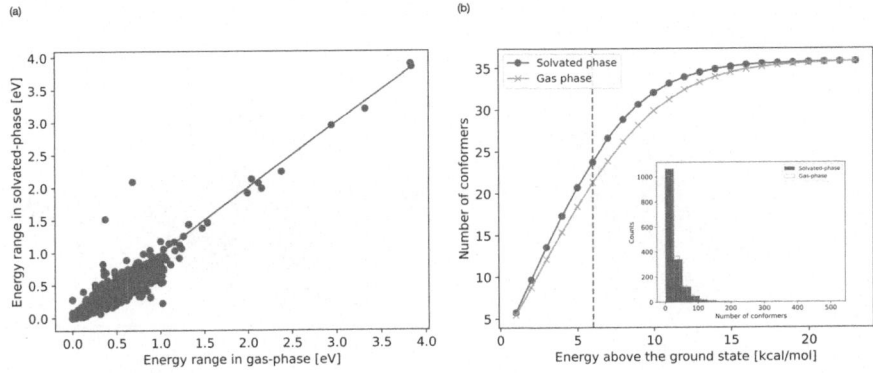

**Fig. 2.** (a) Variation of the energy range for molecules in implicit water and molecules in gas phase. For each unique molecule, the difference between the energy of the highest energy conformer and the lowest energy one was computed. The large majority of molecules lie under the bisect, indicating that generally, the presence of a solvent reduces the energy range. (b) Population of consecutive energy windows above the ground state. The average number of conformers inside windows of increasing energy above the ground state was computed. The solvated phase is consistently more populated than the gas phase in the low-energy regime, indicating a degeneracy.

all molecules in the dataset, defined as the energy difference between the conformers at the highest and lowest energy. The results show a reduction of the energy range, meaning that the same number of conformers is compressed in a smaller energetic window in the solvated phase (see Fig. 2(a)). To see how conformers distribute inside these reduced regions, we counted for each compound the number of conformations inside windows of increased energy from 1 kcal/mol over the ground state to 23 kcal/mol, as shown in Fig. 2(b). Indeed, solvation leads to a degeneration: lower energy windows are systematically more densely populated in the solvated phase than in the gas phase. Accordingly, a direct implication of this finding is that it becomes more challenging to uniquely identify conformations with a specific energy value in a solution, particularly near an energy minimum. Together with this change in energies, it also comes a modification of the radius of gyration, $R_g$, which increases or decreases for molecules with $R_g \approx 5.0$ Å in the gas phase (see Fig.3 in Ref. [11]). This means that the molecular structures can become either more extended or more compact after interacting with the solvent, affecting their flexibility, and hence, the entropy of the system.

To understand how structural changes induced by the implicit solvent model affect the molecular properties, we compared the root mean squared deviation (RMSD) [36–38] between the structure before and after solvation with the variation in the dipole moment $\mu$ (intensive) and many-body dispersion energy $E_{\mathrm{MBD}}$ (extensive). Figure 3 displays the correlation plots for these properties, which appear as unstructured conglomerates of data points. This finding highlights the

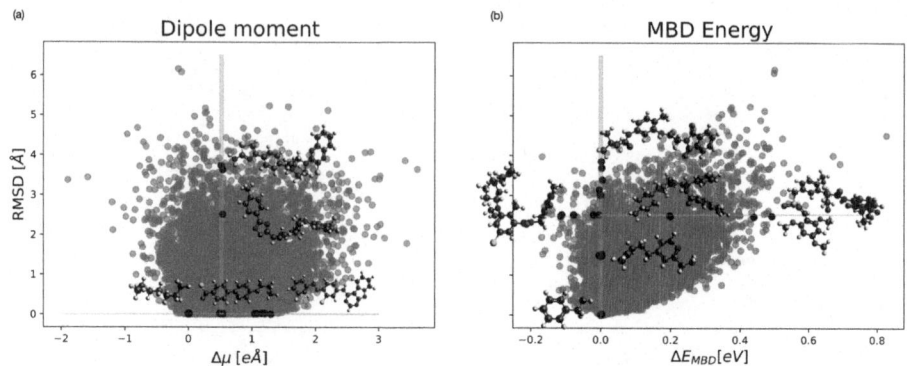

**Fig. 3.** Geometrical change measured as root mean squared deviation (RMSD), as a function of the variation of (a) dipole moment $\Delta\mu$ and (b) many-body dispersion energy $\Delta E_{\mathrm{MBD}}$. Yellow horizontal lines follow the variation of the respective property for molecules showing the same RMSD. Molecules with increasing $\mu$ show a planar geometry, while molecules with increasing $E_{\mathrm{MBD}}$ are associated with structures becoming more compact. On the other hand, yellow vertical lines show the variation of RMSD along molecules with the same property change. The same variation in property is obtainable by a wide variety of compositions—a clear evidence of the "freedom of design" in the chemical space spanned by AQM. (Color figure online)

absence of significant correlations, even when analyzing the variation in properties ($\rho = 0.27$ for $\mu$ and $\rho = 0.68$ for $E_{\mathrm{MBD}}$). $\mu$ exhibits greater variations as the molecular structure becomes more planar, whereas $E_{\mathrm{MBD}}$ shows more significant changes when the structure becomes more compact. Both effects can be rationalized using physics and chemical intuition. Moreover, we found that the "freedom of design" conjecture can be applied in the space defined by the variation of properties under the influence of solvation. In Fig. 3(a,b), it is evident that the same variation in each molecular property can be achieved with diverse RMSD values (indicated by yellow vertical lines), *i.e.*, with molecules that are either more or less flexible. Indeed, it is possible to identify diverse molecules with increasing geometrical differences when considering the mean value of $\Delta\mu = 0.52$ eÅ. For example, $C_{16}H_{14}Cl_2O_2$ does not change its geometry with respect to the gas phase, while $C_{23}H_{23}F_3N_4O_2$ shows an RMSD = 2.5 Å. As an extreme case, we have found molecules with RMSD larger than 3.5 Å, *e.g.*, $C_{23}H_{23}ClN_6S$. Similarly, one can identify molecules for which the change in dispersion energy, $E_{\mathrm{MBD}}$, is zero, but the molecular structure can remain unaltered (as in $C_7H_8O$), or it can have an RMSD > 3 Å ($C_{22}H_{23}N_3O_5$).

As a last analysis, we used the equivariant neural network architecture MACE [32] to learn and predict solvent effects for the QM properties discussed above. We adopted a $\Delta$-learning approach, where the learning targets are $\Delta\mu$ and $\Delta E_{\mathrm{MBD}}$, *i.e.*, we learned corrections to obtain solvated property values from gas-phase ones. The predictions obtained via $\Delta$-learning were compared to the ones obtained by direct learning over four training sets of increasing size, con-

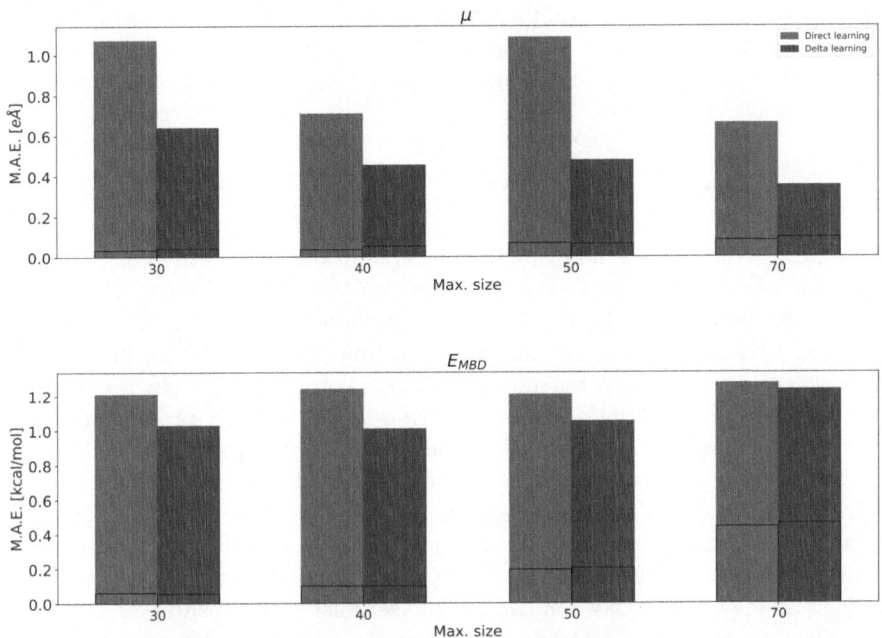

**Fig. 4.** Mean Absolute Error (MAE) for the prediction of dipole moment $\mu$ and many-body dispersion energy $E_{\mathrm{MBD}}$ of solvated molecules using $\Delta$-learning and direct learning methods. Evaluations were computed on molecules larger than 70 atoms, while the training was performed on molecules with a maximum number of atoms of 30, 40, 50, and 70 atoms. The black lines represent the error on the training set.

taining molecules up to 30, 40, 50, and 70 atoms, respectively. The evaluations were always done on molecules larger than 70 atoms. This approach enabled a comprehensive assessment of the robustness of the models across varying molecular sizes. Figure 4 shows the evaluations for the two properties and the four training sets. It is evident how the $\Delta$-learning approach yields better predictions than direct learning for both properties. Moreover, these models display the lowest mean absolute error (MAE) for all training sets—a clear example of their improved scalability. This is more evident when analyzing $\mu$, for which the lowest MAE obtained by $\Delta$-learning is 0.35 e Å for the training set with molecules up to 70 atoms, while MAE of direct learning for the same set is 0.66 e Å. Interestingly, increasing the size of the molecules in the training set does not always increase the accuracy of the predictions. For example, MAE for direct learning in the prediction of $\mu$ increases from 0.71 e Å in the second training set (max. size = 40) to 1.09 e Å in the third training set (max. size = 50). This result could be attributed to the increased complexity of the training data when larger molecules are taken into account, as evidenced by the rise in training error (depicted by the black line). However, further in-depth analysis is necessary to fully elucidate this finding. Similar results were observed for $E_{\text{MBD}}$, although the variance in performance is less pronounced. This is because van der Waals interactions are inherently long-range, and using only a 6 Å cutoff is not enough to correctly describe them.

## 4    Conclusions

In conclusion, our study offers valuable insights into how solvent effects influence the physicochemical properties of an extensive set of large drug-like molecules that, to the best of our knowledge, were previously unknown. These findings could assist the development of machine learning (ML) models for investigating the property space of solvated molecules. For example, the lack of correlation between structural RMSD and the variation of the studied properties showed that solvent can affect the electronic structure of a diverse set of molecules without altering their structure. Hence, ML models should take into consideration the electronic features of these molecules. The matter concerning energetic degeneration also highlights the necessity for a more rigorous molecular description within machine-learned force fields, particularly when applied to systems immersed in a solvent as opposed to those in the gas phase. Lastly, our results on $\Delta$-learning show a way to predict solvated phase properties only using structures and properties of gas-phase molecules. This can be relevant for obtaining information on solvent effects only using the information available in gas-phase, thus avoiding expensive QM calculations. A similar approach can be applied to investigate the dynamics of solvated molecules using data from molecular dynamics simulations [39] or molecular density functional theory [40].

**Acknowledgments.** MH was funded by the European Union's Horizon 2020 research and innovation program under the Marie Sklodowska-Curie Innovative Training Network - European Industrial Doctorate grant agreement No. 956832 "Advanced machine learning for Innovative Drug Discovery". The results discussed in this work were obtained using the computational resources provided by the High Performance Computing (HPC) at the University of Luxembourg. The MACE models were trained on the Luxembourg national supercomputer MeluXina.

**Disclosure of Interests.** The authors have no competing interests to declare that are relevant to the content of this article.

## References

1. Roux, B., Simonson, T.: Implicit solvent models. Biophys. Chem. **78**(1–2), 1–20 (1999)
2. Decherchi, S., Masetti, M., Vyalov, I., Rocchia, W.: Implicit solvent methods for free energy estimation. Eur. J. Med. Chem. **91**, 27–42 (2015)
3. Ringe, S., Oberhofer, H., Hille, C., Matera, S., Reuter, K.: Function-space-based solution scheme for the size-modified poisson-boltzmann equation in full-potential DFT. J. Chem. Theory Comput. **12**(8), 4052–4066 (2016)

4. Klamt, A.: Conductor-like screening model for real solvents: a new approach to the quantitative calculation of solvation phenomena. J. Phys. Chem. **99**(7), 2224–2235 (1995)

5. Onufriev, A.V., Case, D.A.: Generalized born implicit solvent models for biomolecules. Annu. Rev. Biophys. **48**, 275–296 (2019)

6. Xie, L., Liu, H.: The treatment of solvation by a generalized born model and a self-consistent charge-density functional theory-based tight-binding method. J. Comput. Chem. **23**(15), 1404–1415 (2002)

7. Hirata, F.: Molecular theory of solvation, vol. 24. Springer Science & Business Media (2003).https://doi.org/10.1007/1-4020-2590-4

8. Matczak, P., Domagała, M.: Heteroatom and solvent effects on molecular properties of formaldehyde and thioformaldehyde symmetrically disubstituted with heterocyclic groups c 4 h 3 y (where y= o-po). J. Mol. Model. **23**, 1–11 (2017)

9. Odey, M.O., et al.: Unraveling the impact of polar solvation on the molecular geometry, spectroscopy (ft-ir, uv, nmr), reactivity (elf, nbo, homo-lumo) and antiviral inhibitory potential of cissampeline by molecular docking approach. Chem. Phys. Impact **7**, 100346 (2023)

10. Ensing, B., Meijer, E.J., Blöchl, P., Baerends, E.J.: Solvation effects on the SN2 reaction between CH3CL and CL-in water. J. Phys. Chem. A **105**(13), 3300–3310 (2001)

11. Medrano Sandonas, L., et al.: Dataset for quantum-mechanical exploration of conformers and solvent effects in large drug-like molecules. Sci. Data **11**, 742 (2024)

12. Pauletti, M., Rybkin, V.V., Iannuzzi, M.: Subsystem density functional theory augmented by a delta learning approach to achieve kohn-sham accuracy. J. Chem. Theory Comput. **17**(10), 6423–6431 (2021)

13. Ruth, M., Gerbig, D., Schreiner, P.R.: Machine learning of coupled cluster (t)-energy corrections via delta ($\delta$)-learning. J. Chem. Theory Comput. **18**(8), 4846–4855 (2022)

14. Atz, K., Isert, C., Böcker, M.N., Jiménez-Luna, J., Schneider, G.: $\delta$-quantum machine-learning for medicinal chemistry. PCCP **24**(18), 10775–10783 (2022)

15. Ramakrishnan, R., Dral, P.O., Rupp, M., Von Lilienfeld, O.A.: Big data meets quantum chemistry approximations: the $\delta$-machine learning approach. J. Chem. Theory Comput. **11**(5), 2087–2096 (2015)

16. Alibakhshi, A., Hartke, B.: Improved prediction of solvation free energies by machine-learning polarizable continuum solvation model. Nat. Commun. **12**(1), 3584 (2021)

17. Meng, F., Zhang, H., Collins Ramirez, J.S., Ayers, P.W.: Something for nothing: improved solvation free energy prediction with $\delta$-learning. Theor. Chem. Acc. **142**(10), 106 (2023)

18. Lim, H., Jung, Y.: MLsolvA: solvation free energy prediction from pairwise atomistic interactions by machine learning. J. Cheminf. **13**(1), 56 (2021)

19. Gaulton, A., et al.: Chembl: a large-scale bioactivity database for drug discovery. Nucleic Acids Res. **40**(D1), D1100–D1107 (2012)

20. Bickerton, G.R., Paolini, G.V., Besnard, J., Muresan, S., Hopkins, A.L.: Quantifying the chemical beauty of drugs. Nat. Chem. **4**(2), 90–98 (2012)

21. Pracht, P., Bohle, F., Grimme, S.: Automated exploration of the low-energy chemical space with fast quantum chemical methods. PCCP **22**(14), 7169–7192 (2020)

22. Seifert, G., Porezag, D., Frauenheim, T.: Calculations of molecules, clusters, and solids with a simplified LCAO-DFT-LDA scheme. Int. J. Quantum Chem. **58**(2), 185–192 (1996)

23. Elstner, M., et al.: Self-consistent-charge density-functional tight-binding method for simulations of complex materials properties. Phys. Rev. B **58**(11), 7260 (1998)
24. Gaus, M., Cui, Q., Elstner, M.: DFTB3: extension of the self-consistent-charge density-functional tight-binding method (SCC-DFTB). J. Chem. Theory Comput. **7**(4), 931–948 (2011)
25. Tkatchenko, A., DiStasio, R.A., Jr., Car, R., Scheffler, M.: Accurate and efficient method for many-body van der waals interactions. Phys. Rev. Lett. **108**(23), 236402 (2012)
26. Ambrosetti, A., Reilly, A.M., DiStasio, R.A., Tkatchenko, A.: Long-range correlation energy calculated from coupled atomic response functions. J. Chem. Phys. **140**(18) (2014)
27. Stöhr, M., Michelitsch, G.S., Tully, J.C., Reuter, K., Maurer, R.J.: Communication: Charge-population based dispersion interactions for molecules and materials. J. Chem. Phys. **144**(15) (2016)
28. Mortazavi, M., Brandenburg, J.G., Maurer, R.J., Tkatchenko, A.: Structure and stability of molecular crystals with many-body dispersion-inclusive density functional tight binding. J. Phys. Chem. Lett. **9**(2), 399–405 (2018)
29. Perdew, J.P., Ernzerhof, M., Burke, K.: Rationale for mixing exact exchange with density functional approximations. J. Chem. Phys. **105**(22), 9982–9985 (1996)
30. Adamo, C., Barone, V.: Toward reliable density functional methods without adjustable parameters: the pbe0 model. J. Chem. Phys. **110**(13), 6158–6170 (1999)
31. Ringe, S., Oberhofer, H., Reuter, K.: Transferable ionic parameters for first-principles poisson-boltzmann solvation calculations: neutral solutes in aqueous monovalent salt solutions. J. Chem. Phys. **146**(13) (2017)
32. Batatia, I., Kovacs, D.P., Simm, G., Ortner, C., Csányi, G.: MACE: higher order equivariant message passing neural networks for fast and accurate force fields. Adv. Neural. Inf. Process. Syst. **35**, 11423–11436 (2022)
33. Hoja, J., et al.: Qm7-x, a comprehensive dataset of quantum-mechanical properties spanning the chemical space of small organic molecules. Sci. Data **8**(1), 43 (2021)
34. Medrano Sandonas, L., Hoja, J., Ernst, B.G., Vázquez-Mayagoitia, Á., DiStasio, R.A., Tkatchenko, A.: "freedom of design" in chemical compound space: towards rational in silico design of molecules with targeted quantum-mechanical properties. Chem. Sci. **14**(39), 10702–10717 (2023)
35. Góger, S., Medrano Sandonas, L., Müller, C., Tkatchenko, A.: Data-driven tailoring of molecular dipole polarizability and frontier orbital energies in chemical compound space. Phys. Chem. Chem. Phys. **25**, 22211–22222 (2023)
36. Kromann, J.C.: Calculate root-mean-square deviation (rmsd) of two molecules using rotation. Github, Dataset. https://githubcom/charnley/rmsd (2019)
37. Kabsch, W.: A solution for the best rotation to relate two sets of vectors. Acta Crystallogr. Section A: Crystal Phys. Diffr. Theor. Gen. Crystallogr. **32**(5), 922–923 (1976)
38. Walker, M.W., Shao, L., Volz, R.A.: Estimating 3-D location parameters using dual number quaternions. CVGIP: image understanding **54**(3), 358–367 (1991)
39. Böselt, L., Thürlemann, M., Riniker, S.: Machine learning in QM/MM molecular dynamics simulations of condensed-phase systems. J. Chem. Theory Comput. **17**(5), 2641–2658 (2021)
40. Borgis, D., Luukkonen, S., Belloni, L., Jeanmairet, G.: Accurate prediction of hydration free energies and solvation structures using molecular density functional theory with a simple bridge functional. J. Chem. Phys. **155**(2), 024117 (2021)

**Open Access** This chapter is licensed under the terms of the Creative Commons Attribution 4.0 International License (http://creativecommons.org/licenses/by/4.0/), which permits use, sharing, adaptation, distribution and reproduction in any medium or format, as long as you give appropriate credit to the original author(s) and the source, provide a link to the Creative Commons license and indicate if changes were made.

The images or other third party material in this chapter are included in the chapter's Creative Commons license, unless indicated otherwise in a credit line to the material. If material is not included in the chapter's Creative Commons license and your intended use is not permitted by statutory regulation or exceeds the permitted use, you will need to obtain permission directly from the copyright holder.

# Towards Interpretable Models of Chemist Preferences for Human-in-the-Loop Assisted Drug Discovery

Yasmine Nahal[1]([✉]), Markus Heinonen[1], Mikhail Kabeshov[2], Jon Paul Janet[2], Eva Nittinger[3], Ola Engkvist[2,4], and Samuel Kaski[1,5]

[1] Department of Computer Science, Aalto University, Espoo, Finland
{yasmine.nahal,markus.heinonen,samuel.kaski}@aalto.fi
[2] Molecular AI, Discovery Sciences, Biopharmaceuticals R&D, AstraZeneca, Gothenburg, Sweden
{mikhail.kabeshov,jon.janet,ola.engkvist}@astrazeneca.com
[3] Medicinal Chemistry, Research and Early Development, Respiratory and Immunology (R&I), BioPharmaceuticals R&D, AstraZeneca, Gothenburg, Sweden
eva.nittinger@astrazeneca.com
[4] Department of Computer Science and Engineering, Chalmers University of Technology, Gothenburg, Sweden
[5] Department of Computer Science, University of Manchester, Manchester, UK

**Abstract.** In recent years, there has been growing interest in leveraging human preferences for drug discovery to build models that capture chemists' intuition for *de novo* molecular design, lead optimization, and prioritization for experimental validation. However, existing models derived from human preferences in chemistry are often blackboxes, lacking interpretability regarding how humans form their preferences. Enhancing transparency in human-in-the-loop learning is crucial to ensure that such approaches in drug discovery are not unduly affected by subjective bias, noise or inconsistency. Moreover, interpretability can promote the development and use of multi-user models in drug design projects, integrating multiple expert perspectives and insights into multi-objective optimization frameworks for *de novo* molecular design. This also allows for assigning more or less weight to experts based on their knowledge of specific properties. In this paper, we present a methodology for decomposing human preferences based on binary responses (like/dislike) to molecules essentially proposed by generative chemistry models, and inferring interpretable preference models that represent human reasoning. Our approach aims to bridge the gap between human-in-the-loop learning and user model interpretability in drug discovery applications, providing a transparent framework that elucidates how human judgments can shape molecular design outcomes.

**Supplementary Information** The online version contains supplementary material available at https://doi.org/10.1007/978-3-031-72381-0_6.

© The Author(s) 2025

D.-A. Clevert et al. (Eds.): AIDD 2024, LNCS 14894, pp. 58–70, 2025.
https://doi.org/10.1007/978-3-031-72381-0_6

**Keywords:** Human-in-the-loop machine learning · Feature decomposition · User modelling · Interpretability · De novo molecular design

# 1 Introduction

Designing effective molecule scoring functions for drug discovery is a highly complex and multifaceted challenge. This complexity arises from the need to balance multiple objectives, such as potency, selectivity, toxicity and synthetic accessibility (SA), each of which must be optimized simultaneously. Traditional computational methods often struggle to integrate these diverse factors into a single, coherent scoring function, making the discovery process both time-consuming and uncertain. The dynamic nature of biological systems and the vast chemical space further complicate the task, requiring innovative approaches to accurately evaluate molecular efficacy beyond conventional manually-engineered scoring functions.

Human-in-the-loop assisted drug discovery offers a promising solution by incorporating the expertise and intuition of chemists directly into the computational workflow. Unlike static, manually-defined scoring functions, human-in-the-loop learning approaches allow for real-time adjustments based on expert knowledge and evolving insights. This dynamic interaction enables a more nuanced evaluation of candidate molecules, leveraging human judgment to guide the discovery process more effectively. By integrating human expertise, data-driven methods can adapt to new information and improve the relevance and accuracy of molecular predictions.

Several models have been developed to harness human input in drug discovery. Notable among these are the works of Sundin et al. [15] and MolSkill [5]. While Sundin et al. focus on dynamically learning a scoring function from binary human preference responses on proposed designs, MolSkill authors train a neural network using pairwise comparisons between molecules to infer user preferences. These models represent significant advancements in incorporating human preferences into drug discovery, yet they often operate as black-box systems with limited transparency.

The lack of interpretability in these user models is a critical concern. Black-box user models can obscure the rationale behind chemist intuition, making it difficult to trust and validate the outcomes. This may represent a bottleneck to the effective integration of human expertise into the drug discovery process.

A previous study by Kutchukian et al. [10] has shown that medicinal chemists simplify the complex task of identifying promising compounds by focusing on a subset of parameters, despite the complexity involved. Moreover, the study highlighted discrepancies between chemists' reported decision criteria and the actual parameters that influence their choices, emphasizing the need for more transparent and interpretable user models in drug discovery.

To address these challenges, we propose inferring interpretable user models by decomposing observed preference data into meaningful features or molecular descriptors. Feature decomposition [12] is a supervised learning strategy with the potential to efficiently dissect user preference data and understand the underlying factors influencing their decisions. This approach aligns with related work in feature decomposition, which has been successfully applied in various fields such as social science to enhance transparency and robustness of human behavioral models [8]. By adopting a similar strategy, we aim to create interpretable user models of chemist intuition that can be used for molecular design, optimization or experiment prioritization, which can later be integrated in drug discovery pipelines without the need for direct human intervention.

In this paper, we propose a methodology for decomposing human preferences, presented as binary responses (i.e., like/dislike) to molecules either proposed by generative chemistry models or from existing chemical libraries. Our approach seeks to bridge the gap between user modelling and model interpretability for human-in-the-loop assisted drug discovery. By providing a transparent framework, we aim to elucidate how human judgments shape molecular design outcomes, ultimately encouraging the reliance on user models and hybrid machine-user models as scoring functions.

## 2   Methodology

We consider a setting where a user provides a binary response $y \in \{0, 1\}$ to a molecular design $\mathbf{x}$, reflecting their preference for the design. We assume a response model

$$y \sim \text{Ber}\left(\text{sigmoid}\left(\mathbf{w}^T g(\mathbf{x})\right)\right) \tag{1}$$

where the user's binary response comes from a Bernoulli distribution, with the probability given by a sigmoidal function of $\mathbf{w}^T g(\mathbf{x})$. The function $g(\mathbf{x}) \in \mathbb{R}^D$ represents the features considered by the user in their decision, and $\mathbf{w} \in \mathbb{R}^D$ represents a linear weighting of these features. For example, the user might consider the following descriptors

$$g(\mathbf{x}) = (\text{synthetizability}(\mathbf{x}), \text{solubility}(\mathbf{x}), \text{patentability}(\mathbf{x}), \text{activity}(\mathbf{x}))$$

and weigh them by $(0.5, 0.3, 0.4, 0.7)$ respectively. Given that users may make errors in their mental evaluation of descriptors, we assume $g$ is an approximation of some underlying true function $g^\star$

$$g(\mathbf{x}) = g^\star(\mathbf{x}) + \epsilon, \quad \epsilon \sim \mathcal{N}(\mathbf{0}, \text{diag}(\boldsymbol{\sigma}^2)) \tag{2}$$

For instance, a user might misjudge the solubility of a molecule.

Furthermore, we consider that each user has a different labelling function. Let $j$ index the labeller, and define its labelling function as

$$y_j \sim \text{Ber}\left(\text{sigmoid}\left(\mathbf{w}_j^T g_j(\mathbf{x})\right)\right) \tag{3}$$

Thus, by querying a collection of users $(1, \ldots, J)$ about a single molecule, we obtain a set of binary responses $(y_1, \ldots, y_J)$, each stemming from different preference functions $\mathbf{w}_j^T g_j(\mathbf{x})$.

We assume a dataset of binary 'votes' $\mathbf{Y} = (y_{ij}) \in \mathbb{R}^{N \times J}$, where $y_{ij}$ is the label of the $i$-th molecule from the $j$-th user.

Our goal is to infer the mental variables $\{\mathbf{w}_j, g_j\}$, but this task is unrealistic in its current form, since we have no direct knowledge of the mental descriptors $g_j$ or the weightings $\mathbf{w}_j$ of the users.

Instead, we simplify the problem by assuming that all experts share the same descriptors. More precisely, we assume that the set of descriptors used by any expert is the union of all expert descriptors, with unused descriptors showing as zeros in $\mathbf{w}$. We thus reformulate the model as

$$y_j \sim \text{Ber}\left(\text{sigmoid}\left(\mathbf{w}_j^T g(\mathbf{x})\right)\right) \tag{4}$$

Given that fitting binary outcome variables directly might not yield a closed-form solution, we consider using a probit link function instead of the logit (sigmoid) function. The probit link can simplify the sampling process, resulting in a straightforward Gibbs sampler when only the weights are learned. This modification is expressed as:

$$y_j \sim \text{Ber}\left(\Phi\left(\mathbf{w}_j^T g(\mathbf{x})\right)\right) \tag{5}$$

where $\Phi$ denotes the cumulative distribution function (CDF) of the standard normal distribution.

Finally, our problem is to infer the posterior

$$p(\mathbf{W}, g \mid \mathbf{Y}) \propto p(\mathbf{Y} \mid \mathbf{W}, g) p(g) \prod_j p(\mathbf{w}_j) \tag{6}$$

where all weights $\mathbf{w}_j$ share the same prior. We aim to determine the mental descriptors $g$ and the user preferences $\mathbf{W} = (\mathbf{w}_1, \ldots, \mathbf{w}_J) \in \mathbb{R}^{J \times D}$. Prior knowledge can be incorporated into this problem. First, the preferences $\mathbf{w}_j$ are assumed to be sparse, for instance, by employing a Horseshoe prior $p(\mathbf{w}_j) = \mathcal{HS}(\mathbf{w}_j)$. Second, the descriptor function $g$ can be fixed to a dictionary of known descriptors (e.g., from chemoinformatics software) or can be the feature embedding of a molecular deep learning network (e.g., MolBert). A more advanced approach would be to treat these as the prior mean functions and infer slight fine-tuning of them.

Given the expense of human evaluations, even if this model does not have a closed-form solution, MCMC methods such as Hamiltonian Monte Carlo (HMC) are likely feasible options. These methods can handle the complex posterior distributions involved in our problem and provide robust estimates of the parameters.

In summary, while a closed-form solution would be ideal for computational simplicity, the use of probabilistic methods like HMC offers a practical and effective route for inference in our setting. We will implement and compare both logit and probit models, evaluating their performance using real-world data to ensure the robustness and applicability of our approach.

# 3   Experiments

We conducted experiments to infer interpretable user models for chemist preferences in molecular design using Bayesian inference with a Stan model. Our approach involved querying experts to provide binary preference responses (*like/dislike*) for a dataset of molecules represented by molecular descriptors. The model assumes that each expert's preference is influenced by a weighted combination of these descriptors.

## 3.1   Experimental Setup

**Data Collection.** We collected a dataset consisting of binary responses ($\mathbf{Y}$) from $J = 3$ experts for $N = 150$ molecules generated using the molecular design tool REINVENT [3]. The experts were asked to rate those molecules based on how much they align with the molecular design objective of producing novel binders for the Dopamine receptor D2 (DRD2). Feedback from experts was collected in real-time through the Metis interface [9] by sampling (after a defined number of reinforcement learning steps) molecules from the generative chemistry model implemented in REINVENT. This generative chemistry model was trained to maximize predicted DRD2 probabilities by a Quantitative Structure-Activity Relationship (QSAR) model. A screenshot of the interface is provided in Figure S6, where structures of generated molecules are displayed alongside their DRD2 activity probabilities. The experts were asked to express, on a scale from 0 to 100, how much they liked the proposed DRD2 binders. Expert scores were transformed into binary labels using a threshold value of 50 and included in the initial training dataset of the DRD2 QSAR model, which was then used to guide the generation of subsequent DRD2 binders by REINVENT. This iterative process ensured that the generative chemistry model continually improved in alignment with expert preferences. All three participating experts are co-authors of this manuscript.

For the set of evaluated molecules, we calculated molecular descriptors using RDKit [1], which include the molecular weight (MolWt), number of rotatable bonds (NumRotaBonds), the logarithm of the octanol-water partition coefficient or LogP (MolLogP), the number of aromatic rings (NumAromRings), the number of hydrogen bond acceptors (HBA) and donors (HBD), the topological surface area (TPSA), and the structural alerts or undesirable substructures according to the Quantitative Estimate of Drug-likeness (QEDAlerts) [2]. For the latter, we modified the standard QED implementation in RDKit by setting the weights for all other properties (MolWt, MolLogP, HBA, HBD, TPSA, NumRotaBonds, NumAromRings) to 0 and only keeping the weight for the presence of undesirable substructures to 1. This ensures that the QED score solely reflects the presence of structural alerts. Additionally, we used the SA score developed by Ertl et al. [7], as well as the probability of DRD2 bioactivity according to the classifier developed by Olivecrona et al. [13], as descriptors that can explain the user preference responses.

We analyzed the Pearson correlations among the molecular descriptors used for this study (Figure S7). Notably, MolLogP shows the strongest positive correlation (0.78) with TPSA. All correlations, ranging from -0.63 to 0.78, are indicative of meaningful relationships between descriptors that can enhance model accuracy and interpretability.

**Model Specification.** The Bayesian model was implemented using the Stan probabilistic programming language [4]. The model included:

- Parameters:
  - $\tau$: Global scale parameter controlling the overall sparsity of weights assigned to the molecular descriptors.
  - $\lambda_j$: Local scale parameters for each expert $j$.
  - **w**: Preference weights matrix, where each column represented the weights for one expert across all descriptors.
- Priors:
  - $\tau \sim \text{Cauchy}(0, \tau_0)$: Cauchy prior for global shrinkage.
  - $\lambda_j \sim \text{Cauchy}(0, 1)$: Cauchy priors for local shrinkage.
  - $\mathbf{w} \sim \text{Normal}(0, \lambda_j \cdot \tau)$: Normal priors for weights adjusted by local scales.
- Likelihood:
  - $\mathbf{Y}_{nj} \sim \text{Bernoulli}(\text{logit}(\mathbf{X} \cdot \mathbf{w}_{.j}))$: Likelihood of expert $j$'s response based on the linear combination of molecular descriptors weighted by $\mathbf{w}_{.j}$.

## 3.2 Implementation

The Stan model was compiled and fitted to the data using Hamiltonian Monte Carlo (HMC) sampling (2000 iterations, 2 Markov chains with a maximum tree depth of 15 and the parameter adapt_delta set to 0.99). This approach enabled us to approximate the posterior distribution of parameters **w** and $\lambda_j$, which represent the preference weights and local scale parameters, respectively.

Since the dataset is already very small (due to limited resource availability for human data collection), we die not split it into training and testing, and chose to fit the model to the entire dataset instead to reach the highest accuracy.

Convergence diagnostics, including the $\hat{R}$ statistic and the trace plots, were performed to assess the model's convergence and ensure reliable inference across multiple chains. The $\hat{R}$ statistic, also known as the potential scale reduction factor, is a convergence diagnostic used to assess whether the Markov chains in the MCMC sampling have converged to the target distribution (i.e., response labels). Specifically, it compares the variance within chains to the variance between chains. A value close to 1 (typically $\hat{R} < 1.1$) indicates convergence, which is what we have observed with our model. Trace plots are visual representations that show how the Markov chain samples evolve over iterations, allowing to diagnose issues like non-stationarity and mixing problems. Our trace plots (Figure S5) show that the Markov chains have mixed properly (low divergence).

## 3.3   Benchmark

We compared our model against a non-probabilistic logistic regression (LogReg) and a Random Forest Classifier (RFC), implemented using the Scikit-learn package [14]. The purpose is to demonstrate that our model is more transparent than its non-probabilistic counterparts, allowing for direct interpretability of the reasoning process behind the human preferences, in addition to a reasonable classification accuracy. The same set of molecular descriptors described in Sect. 3.1 was used to train the LogReg and RFC models on the 150 human-rated molecules by each expert, individually. The classification accuracy scores (i.e., percentages of correctly classified molecules into liked or disliked) were calculated for each individual user model, then the average accuracy scores were reported. To assess the interpretability of the RFC models, Shapely values for tree-based algorithms were computed [11]. For LogReg models, we analyzed feature importance.

# 4   Results

## 4.1   Interpretability of Human Preferences

We consider that a model is able to accurately interpret human preferences based on how the participating experts described their reasoning. Our Stan model effectively deciphered the human reasoning behind the preference dataset for the DRD2 binders. The learned descriptor contributions are weights are illustrated in Fig. 1.

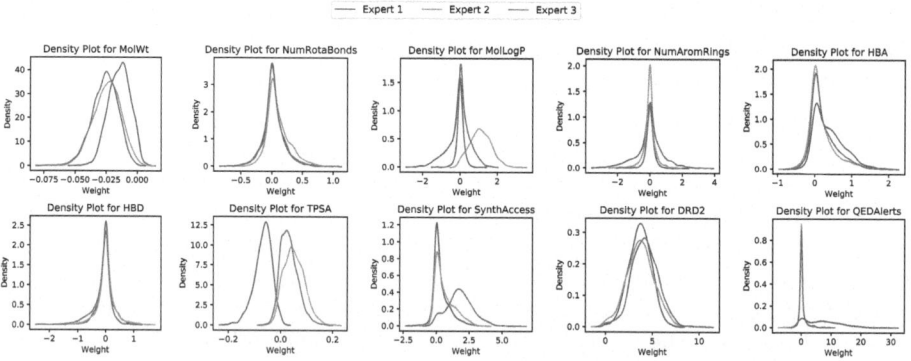

**Fig. 1. Density plots of the molecular descriptor weights learned by the Stan model for each expert.** Each subplot corresponds to a different descriptor, with density curves representing the weight distributions for Expert 1 (blue), Expert 2 (orange) and Expert 3 (green). These plots illustrate the variations in how each expert weighted the descriptors, reflecting their individual preferences and reasoning. Notably, the plots show that the DRD2 bioactivity descriptor was consistently important across all experts, while other descriptors such as MolLogP, SynthAccess and QEDAlerts had varying levels of importance depending on the expert. (Color figure online)

When asked to describe their personal experiences from interacting with DRD2 binders generated using REINVENT, Expert 1 highlighted their focus on the structural characteristics of the generated molecules, rejecting many at the initial stages of the interaction because they appeared too different and "odd" compared to known DRD2 actives. This was accurately captured by the Stan model, showing greater correlation between Expert 1 preferential feedback and the QEDAlerts descriptor (Fig. 2). The focus of Expert 1 on the presence of structural alerts in the generated DRD2 binders has led to the generation of more drug-like molecules, which can be observed through an increased QED score (Table S1 where molecular generation performance metrics are reported for top-scoring DRD2 binders generated by REINVENT after incorporating expert preference feedback).

Conversely, Expert 3 described that they were more concerned with the SA of the generated DRD2 binders. This preference was well captured by the higher estimated weights for the SynthAccess descriptor (Fig. 1) and correlation between Expert 2 preferential feedback and SynthAccess (Fig. 2). Expert 2 described that they rated the molecules based on how much they liked them as a lead, aiming to select molecules that would be synthesizable, stable and with reasonable lipophilicity to maximize their chance for being made and tested. The Stan model's learned weights revealed that Expert 2 indeed prioritized the molecular LogP followed by synthetic accessibility, as indicated by the higher weights for those descriptors (Fig. 1) and stronger correlation with MolLogP (Fig. 2). Moreover, a higher percentage of lead-like compounds according to the rule of three (RO3) [6] for molecular LogP was identified based on feedback from Expert 2 (Table S1). Expert 2 showed similarities with Expert 3 in their reasoning regarding DRD2 binders: they both acknowledged not having any particular knowledge of the target or known binders.

Interestingly, the weights for the DRD2 bioactivity descriptor were high for all three experts, indicating that the model successfully captured that the preference feedback was related to the rating of DRD2 binders. These findings are consistent with the descriptions provided by the experts upon the completion of the interaction exercise, validating the model's ability to interpret and reflect their reasoning accurately.

The interpretability analysis from the RFC models also highlighted the importance of the DRD2 activity descriptor in explaining user preference feedback 3. For Expert 1, the RFC models accurately captured their preference for more complex molecular structures but did not fully reflect their reliance on the presence of structural alerts that could undermine drug likeness. For Experts 2 and 3, MolLogP and SynthAccess were correctly identified as important descriptors in explaining their feedback. However, MolWt was also identified as significantly important, though it was not explicitly emphasized in the expert feedback. Therefore, we consider the RFC models' interpretations to be close to the Stan model's performance, with the latter being the most aligned with the expert descriptions Fig. 3.

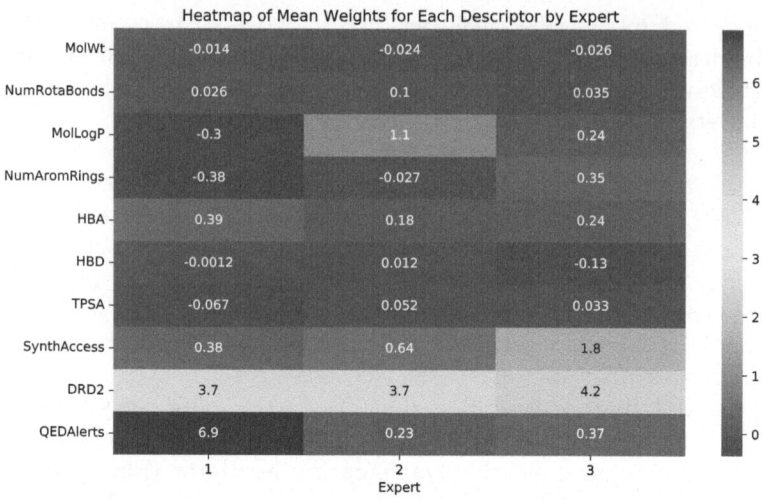

**Fig. 2. Correlation between each expert and molecular descriptors according to the fitted Stan model.** Heatmap matrix showing the relationship between each expert's feedback and the molecular descriptors used for fitting the Stan model. The matrix highlights a higher correlation between Expert 1 and `QEDAlerts`, Expert 2 and `MolLogP`, and Expert 3 and `SynthAccess`. All experts show a high correlation with the DRD2 activity descriptor, indicating its importance to explain the expert preferential responses.

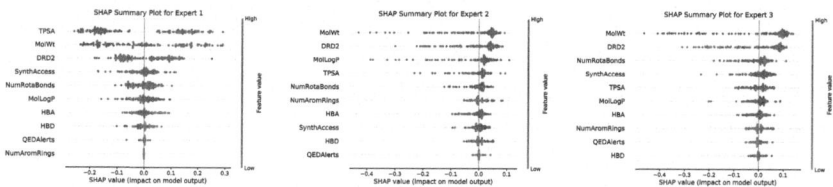

**Fig. 3. SHAP summary plots for the Random Forest Classifier predictions**, illustrating the importance of various molecular descriptors for each expert. The plots provide a visual explanation of how each descriptor contributes to the model's predictions, with distinct patterns emerging for each expert that align with their feedback preferences.

Conversely, the LogReg models provided the least accurate interpretation of descriptor importance. They failed to clearly capture Expert 1's emphasis on structural alerts and Expert 3's focus on SA for the generated DRD2 binders Fig. 4. This suggests that while the LogReg models can provide some insights, they are not as reliable as the Stan and RFC models in reflecting the experts' reasoning.

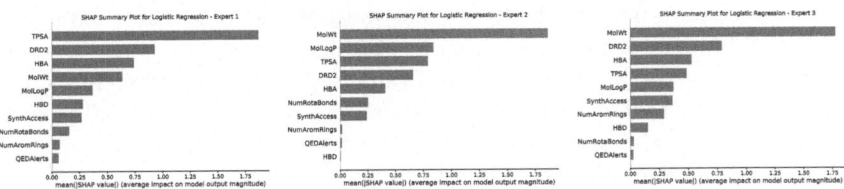

**Fig. 4. SHAP summary plots for the Logistic Regression predictions**, depicting the significance of molecular descriptors in determining the experts' feedback. These plots highlight the differences in descriptor importance across experts and provide insight into the Logistic Regression model's interpretation of the data.

## 4.2 Accuracy in Predicting Human Preferences

We compared our model against a non-probabilistic logistic regression (LogReg) and a Random Forest Classifier (RFC), implemented using the Scikit-learn package [14]. The primary goal of this comparison is to demonstrate the interpretability of our model in contrast to its non-probabilistic counterparts, while also showcasing its reasonable classification accuracy. The same molecular descriptors were used to train the LogReg and RFC models on the 150 human-rated molecules by each expert, individually. The classification accuracy scores (i.e., percentages of correctly classified molecules into liked or disliked) were calculated for each individual user model, then the average accuracy scores were reported.

Despite the slight edge in predictive performance by the RFC model, the Stan (Bayesian) model offers significant advantages in terms of interpretability. Unlike its non-probabilistic counterparts, the Stan model provides posterior distributions for the learned weights of molecular descriptors. This feature not only allows for a clear understanding of the importance of different descriptors but also incorporates the uncertainties associated with these weights. Such probabilistic insights are crucial for gaining a deeper understanding of the factors driving experts' preferences and ensuring that the model's predictions are not only accurate but also comprehensible and justifiable.

In summary, while the RFC model boasts the highest predictive accuracy, the Stan model's interpretability and ability to quantify uncertainties make it a valuable tool for elucidating the rationale behind experts' preferences in molecular design. This dual benefit of accuracy and interpretability underscores the potential of Bayesian models in explaining complex decision-making processes such as human rating.

## 5 Discussion

In this work, we developed and evaluated models to decipher and predict human preferences in molecular design, focusing on the interpretation of these preferences using various, known and self-explanatory molecular descriptors. The Stan (Bayesian) model, Logistic Regression (LogReg), and Random Forest Classifier

(RFC) were employed to fit the preference data provided by three experts on a set of DRD2 binders proposed by a molecular design tool.

The posterior distributions of the Stan model provided insights into the importance of different molecular descriptors for each expert, revealing distinct patterns in their preferences. Expert 1 focused on the structural complexity of molecules and presence of undesired structures for drug-likeness, as evidenced by higher correlations with QED structural alerts. Expert 2's preferences indicated a focus on lipophilicity and synthetic accessibility, similar to Expert 3. Interestingly, all three experts were characterized by high correlations with the DRD2 activity descriptor, aligning with the core objective of the feedback exercise which is to rate DRD2 binders. Notably, the interpretations derived from the Stan model aligned the closest with the reasoning process described by the experts themselves, enhancing the model's ability to accurately explain the expert preference data and decision-making.

The interpretability analysis from the RFC models also highlighted the importance of the DRD2 activity descriptor. For Expert 1, the RFC model did not capture the reliance on structural alerts. For Experts 2 and 3, the molecular LogP and synthetic accessibility descriptors were correctly identified as important, although the RFC model also highlighted molecular weight as a significant factor, which was not explicitly mentioned in expert descriptions. The LogReg models, however, provided a less accurate interpretation.

In terms of predictive accuracy, the RFC model achieved the highest performance, followed by the Stan model and the LogReg model. Despite the superior predictive accuracy of the RFC model, the Stan model's ability to better capture the relationships between the human reasoning processes and the molecular descriptors, and to quantify uncertainties through the posterior distributions, makes it a more interpretable and insightful tool for understanding the reasoning behind experts' preferences.

One of the main limitations of this study is the small amount of expert preference data available. This limited data set may not fully capture the variability and complexity of experts' decision-making processes. Consequently, the generalizability of our models to new, unseen data remains an open question. Future work should focus on collecting more extensive preference data from a larger and more diverse group of experts. This would not only improve the robustness and generalizability of our model but also provide a more comprehensive understanding of how different molecular descriptors influence human preferences in molecular design.

Additionally, it would be valuable to validate the model on unseen data to assess their predictive performance in real-world scenarios. This validation step is crucial for ensuring their practical applicability in molecular design tasks.

In conclusion, while our preliminary model demonstrate high predictive accuracy and provide valuable insights into the reasoning behind experts' preferences, addressing the limitations related to data size and generalizability are essential steps for future work.

**Acknowledgments.** This study was partially funded by the European Union's Horizon 2020 research and innovation program under the Marie Sklodowska-Curie Innovative Training Network European Industrial Doctorate grant agreement No. 956832 "Advanced Machine Learning for Innovative Drug Discovery". Further, this work was supported by the Academy of Finland Flagship program: the Finnish Center for Artificial Intelligence FCAI, and the UKRI Turing AI World-Leading Researcher Fellowship, [EP/W002973/1].

**Disclosure of Interests.** The authors have no competing interests to declare that are relevant to the content of this article.

# References

1. RDKit: open-source cheminformatics. https://www.rdkit.org
2. Bickerton, G.R., Paolini, G.V., Besnard, J., Muresan, S., Hopkins, A.L.: Quantifying the chemical beauty of drugs. Nat. Chem. **4**(2), 90–98 (2012). https://doi.org/10.1038/NCHEM.1243
3. Blaschke, T., et al.: REINVENT 2.0: an AI tool for de novo drug design. J. Chem. Inf. Model. **60**(12), 5918–5922 (2020)
4. Carpenter, B., et al.: Stan: a probabilistic programming language. J. stat. softw. **76**(1) (2017)
5. Choung, O.H., Vianello, R., Segler, M., Stiefl, N., Jiménez-Luna, J.: Learning chemical intuition from humans in the loop. ChemRxiv (2023). https://doi.org/10.26434/chemrxiv-2023-knwnv
6. Congreve, M., Carr, R., Murray, C., Jhoti, H.: A 'rule of three' for fragment-based lead discovery? Drug discovery today **8**(19), 876–877 (2003). https://doi.org/10.1016/S1359-6446(03)02831-9, https://www.sciencedirect.com/science/article/pii/S1359644603028319
7. Ertl, P., Schuffenhauer, A.: Estimation of synthetic accessibility score of drug-like molecules based on molecular complexity and fragment contributions. Journal of cheminformatics **1**, 1–11 (2009)
8. Fennell, P.G., Zuo, Z., Lerman, K.: Predicting and explaining behavioral data with structured feature space decomposition. EPJ Data Sci. **8**(1), 1–27 (2019)
9. Janosch, M., Nahal, Y., Jannik Bjerrum, E., Kabeshov, M., Engkvist, O., Kaski, S.: A python-based user interface to collect expert feedback for generative chemistry models. ChemRxiv. 2024; https://doi.org/10.26434/chemrxiv-2024-zs5xp This content is a preprint and has not been peer-reviewed (2024)
10. Kutchukian, P.S., et al.: Inside the mind of a medicinal chemist: the role of human bias in compound prioritization during drug discovery. PLoS ONE **7**(11), e48476 (2012)
11. Lundberg, S., Lee, S.I.: A unified approach to interpreting model predictions (2017). https://arxiv.org/abs/1705.07874
12. Maimon, O., Rokach, L.: Improving supervised learning by feature decomposition. In: Eiter, T., Schewe, K.D. (eds.) Foundations of Information and Knowledge Systems, pp. 178–196. Springer, Berlin Heidelberg, Berlin, Heidelberg (2002). https://doi.org/10.1007/3-540-45758-5_12
13. Olivecrona, M., Blaschke, T., Engkvist, O., Chen, H.: Molecular de-novo design through deep reinforcement learning. J. Cheminformatics **9**(1), 48 (2017). https://doi.org/10.1186/s13321-017-0235-x

14. Pedregosa, F., et al.: SciKit-learn: machine learning in python. J. Mach. Learn. Res. **12**, 2825–2830 (2011)
15. Sundin, I., et al.: Human-in-the-loop assisted de novo molecular design. J. Cheminformatics **14**(1), 86 (2022). https://doi.org/10.1186/s13321-022-00667-8

**Open Access** This chapter is licensed under the terms of the Creative Commons Attribution 4.0 International License (http://creativecommons.org/licenses/by/4.0/), which permits use, sharing, adaptation, distribution and reproduction in any medium or format, as long as you give appropriate credit to the original author(s) and the source, provide a link to the Creative Commons license and indicate if changes were made.

The images or other third party material in this chapter are included in the chapter's Creative Commons license, unless indicated otherwise in a credit line to the material. If material is not included in the chapter's Creative Commons license and your intended use is not permitted by statutory regulation or exceeds the permitted use, you will need to obtain permission directly from the copyright holder.

# Atom-Level Quantum Pretraining Enhances the Spectral Perception of Molecular Graphs in Graphormer

Alessio Fallani[1,3(✉)], José Arjona-Medina[1], Konstantin Chernichenko[1],
Ramil Nugmanov[1], Jörg Kurt Wegner[2], and Alexandre Tkatchenko[3]

[1] Drug Discovery Data Sciences, Janssen Pharmaceutica NV, Turnhoutseweg 30,
2340 Beerse, Belgium
[2] Johnson & Johnson Innovative Medicine, 301 Binney Street, Cambridge,
MA 02142, USA
[3] Department of Physics and Materials Science, University of Luxembourg,
1511 Luxembourg City, Luxembourg
alessio.fallani.001@student.uni.lu

**Abstract.** This study explores the impact of pretraining Graph Transformers using atom-level quantum-mechanical features for molecular property modeling. We utilize the ADMET Therapeutic Data Commons datasets to evaluate the benefits of this approach. Our results show that pretraining on quantum atomic properties improves the performance of the Graphormer model. We conduct comparisons with two other pretraining strategies: one based on molecular quantum properties (specifically the HOMO-LUMO gap) and another using a self-supervised atom masking technique. Additionally, we employ a spectral analysis of Attention Rollout matrices to understand the underlying reasons for these performance enhancements. Our findings suggest that models pretrained on atom-level quantum mechanics are better at capturing low-frequency Laplacian eigenmodes from the molecular graphs, which correlates with improved outcomes on most evaluated downstream tasks, as measured by our custom metric.

## 1 Introduction

In recent years, the application of deep learning techniques has brought about a paradigm shift in molecular representation learning, playing a pivotal role in a wide array of biochemical endeavors including property modeling and drug design [3,5–7,18,20]. Leveraging deep learning methodologies has enabled researchers to extract intricate features from molecular data, thereby enhancing our understanding of molecular structures and their interactions. However, despite the remarkable successes achieved, challenges such as data scarcity and generalizability remain pertinent concerns in the field [3,4,8,10,12]. To address these challenges, the concept of pretraining models on related tasks or employing self-supervised learning strategies has gained significant traction. Pretraining serves

© The Author(s) 2025
D.-A. Clevert et al. (Eds.): AIDD 2024, LNCS 14894, pp. 71–81, 2025.
https://doi.org/10.1007/978-3-031-72381-0_7

$$|a_0| \geq |a_1| \geq |a_2| \geq ... \geq |a_{N-1}|$$

$$\tilde{A} = a_0 |a_0\rangle \langle a_0| + a_1 |a_1\rangle \langle a_1| + a_2 |a_2\rangle \langle a_2| + ... + a_{N-1} |a_{N-1}\rangle \langle a_{N-1}|$$

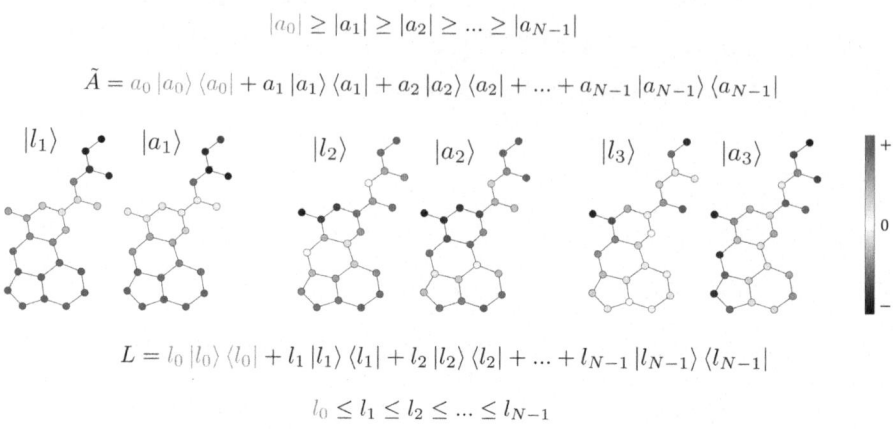

$$L = l_0 |l_0\rangle \langle l_0| + l_1 |l_1\rangle \langle l_1| + l_2 |l_2\rangle \langle l_2| + ... + l_{N-1} |l_{N-1}\rangle \langle l_{N-1}|$$

$$l_0 \leq l_1 \leq l_2 \leq ... \leq l_{N-1}$$

**Fig. 1.** Example of the similarity observed between the eigenvectors $|a_i\rangle$ of the rollout matrix $\tilde{A}$ and the eigenvectors $|l_i\rangle$ of the graph Laplacian $L$ for a Graphormer pretrained on atomic QM properties. The molecule used here comes from the TDC dataset.

as a means to provide models with a foundational understanding of molecular structures, enabling them to learn meaningful representations even in the presence of limited or noisy labeled data. By leveraging pretraining techniques, researchers aim to enhance model generalizability and performance across a spectrum of downstream tasks [15, 19, 25–27].

In this context, our study focuses on investigating the impact of pretraining on atom-level quantum mechanical (QM) properties, associated with fundamental aspect of molecular behavior with profound implications in biochemical research [2] and present in an incresing number of public datasets [14, 17, 21], implemented on Graphormer neural network [28], an instance of the increasingly popular family of Graph Transformer (GT) architectures [22]. Specifically, we compared the efficacy of such pretraining with alternative strategies such as pretraining on a molecular quantum property (HOMO-LUMO gap) and masking, an atom-level self-supervised pretraining method. As downstream tasks that are relevant for applications in the pharmaceutical industry we utilized the ADMET properties dataset from the Therapeutics Data Commons (TDC) [16]. For each pretraining technique and downstream property we compared the model performance with a spectral analysis of the Attention Rollout matrix, to understand in approximation the contributing factors to the model. This analysis reveals that models pretrained with atom-level quantum properties and with masking extract graph spectral properties in the form of Laplacian eigenvectors. Moreover, we observe that models pretrained with atom-level quantum properties can extract more low-frequency Laplacian eigenmodes from the input graph signal and we demonstrate how this effect correlates with improved performances on a good part of the downstream tasks.

## 2    Methods

We consider a custom implementation of Graphormer [24, 28] as an instance of network that belongs to the category of GTs. As baseline we employed the non-pretrained Graphormer version, which was compared with pretrained models for a total of 8 different cases: one per each of the 4 atom-resolved QM properties (atomic charges, NMR shielding constants, electrophilic and nucleophilic Fukui function indexes), one considering all atomic properties in a multi-task setting, one for the considered molecular property (HOMO-LUMO gap), and one for masking node pretraining. A spectral analysis of the Attention Rollout matrix $\tilde{A}$ is then performed to gain insights into the behaviour of each obtained model.

**Model Description.** Graphormer is a GT where the input molecule is seen as a graph where atoms are nodes and bonds are edges. This model in general works by encoding the atoms in the molecule tokenized based on their atom type, and then repeatedly applying self-attention layers with an internal bias term before the softmax. This term is based on the topological distance matrix of the molecular graph and allows to encode the structural information of the molecular graph. In particular, the network employed in this work is an implementation of Graphormer from [24], inspired by the implementation from [28]. In this implementation the centrality encoder is adapted from using only explicit neighbours to including both explicit atoms and implicit hydrogens. As a result of the combination of this modified centrality encoding together with the usual atom type encoder, the hybridization of atoms is handled implicitly. For this reason this implementation does not present any edge encoder component.

**Datasets.** For pretraining, we used a publicly available dataset [13] consisting of 136k organic molecules and containing, among other things, atomic properties calculated with quantum chemistry methods. Each molecule is represented by a single conformer generated using the Merck Molecular Force Field (MMFF94s) in RDKit library. The initial geometry for the lowest-lying conformer was then optimized at the GFN2-xtb level of theory followed by refinement of the electronic structure with DFT (B3LYP/def2svp). Notice that while the 3D structure is used for the computation of the properties, this is not used in the model where the molecule is represented as 2D input (graph). The advantage of the described dataset is several reported atomic properties: charge, electrophilicity, and nucleophilicity Fukui indexes and an NMR shielding constant. The same set of molecules was used for masked node pretraining. Another pretraining dataset, PCQM4Mv2, consists of a single molecular property per molecule, a HOMO-LUMO gap that was also calculated using quantum chemistry methods https://ogb.stanford.edu/docs/lsc/pcqm4mv2/. It was curated under the PubChemQC project [23]. For the benchmarking of the obtained pretrained models, we used the absorption, distribution, metabolism, excretion, and toxicity (ADMET) group of the TDC dataset, consisting of 9 regression and 13 binary classification tasks for modeling biochemical molecular properties https://tdcommons.ai/benchmark/admet_group/overview/.

**Atom-Level Quantum Pretraining.** The pretraining on atom-level quantum mechanical properties is achieved via regression task. In the model, each node corresponds to a heavy (non-hydrogen) atom. Accordingly, the obtained node embeddings, are used to train atom-level properties via a linear layer. The model is trained on the dataset from [13] on each one of the available atomic properties, as well as on all of them at the same time in a multi-task setting. As a result, we obtain from this pretraining 5 different pretrained models. In each case except for HOMO-LUMO gap the model was trained as a regression task using L1 loss. A batch size of 100 was used with a fixed learning rate of $10^{-4}$. In the case of HOMO-LUMO pretraining a triangular cyclic scheduling was employed with a minimum value of $2 \times 10^{-5}$ and a maximum value of $2 \times 10^{-4}$. The training was stopped using an early stopping criterion with patience of 100 epochs. For what concerns labels, the properties were not scaled except for a constant scaling factor of $10^{-2}$ for NMR shielding constants as we observed it to helped convergence.

**Molecule-Level Quantum Pretraining.** The pretraining on molecular quantum properties is achieved via a simple regression task where the output is obtained by applying a linear layer to the class token embedding at the last layer of the network. The model is trained on the modeling of HOMO-LUMO gap on the PCQM4Mv2 dataset. We used the same training hyperparameters as the ones indicated in 2. As a result of this pretraining we obtain an additional pretrained model to consider for the downstream tasks.

**Masking.** Masking pretraining is carried out in a similar way to what is usually done in BERT-based models [9,11]. This procedure entails randomly masking 15% of the input graph node tokens by replacing them with the mask token, and then training the model to restore the correct node type from the masked embedding as a multi-class classification task. The model is trained on the molecular structures present in the dataset used for atomic QM properties. As a result, we obtain one additional pretrained model to consider for the downstream tasks. The hyperparmeters used for this pretraining are the same as the ones used in 2, while the loss employed is a cross entropy loss.

**Downstream Tasks.** The training and testing on downstream tasks is carried out on the ADMET group from the TDC dataset in the same way as any molecular property modeling. For splittings and evaluation metrics we follow the guidelines of the benchmark group that we consider, hence we refer to [16]. The pretrained models are fine tuned for each downstream task by training without freezing any layer. Additionally, we also train a model from scratch, obtaining a total of 8 final models per each of the 5 default train/validation splitting seeds on each task (considering 22 tasks, 5 seeds and 8 models we obtain a total of

880 models). The hyperparameters used in each downstream task are the same: the batch size used is 32, while for what concerns the learning rate a triangular cyclic scheduling was employed with a minimum value of $2 \times 10^{-5}$ and a maximum value of $2 \times 10^{-4}$. The training is stopped with an early stopping criterion with patience of 200. The loss used for regression tasks is L1 loss, while for classification tasks a censored regression approach is used using again L1 loss with right censor set at 0 for negative examples and left censor set at 1 for positive examples. For what concerns regression labels, given the diversity of the tasks we opted for a standard scaling. Finally, the performances on each task's test set are obtained per each pretraining case by taking mean and standard deviation of the performances obtained by the 5 models coming from the 5 different training/validation splits.

**Spectral Analysis of Attention Rollout.** To have a better understanding of the mechanism behind the pretrained models' improvements, we shift our focus on the analysis of attention weights. What we aim to understand is along which directions the input molecular representation is decomposed when passed through a given model. In order to do so we start by considering the Attention Rollout matrix [1] $\tilde{A}$ as a proxy for the model's action on the input. While this approximation is a strong one, as we will see it provides a number of non-trivial insights. For the definition of $\tilde{A}$ we refer to [1]. We start by considering a simple spectral decomposition of $\tilde{A}$ (from here on we will make use of the bra-ket notation):

$$\tilde{A} = \sum_{i=0}^{N-1} a_i |a_i\rangle \langle a_i| \tag{1}$$

with $a_i \in \mathbb{C}$ and $|a_0| \geq |a_1| \geq ... \geq |a_{N-1}|$ and, based on an empirical observation on one of the pretrained Graphormers (see Fig. 1), we analyse the similarity of the eigenvectors $|a_i\rangle$ with the eigenvectors of the Laplacian matrix $L$ of the input molecular graph decomposed as

$$L = \sum_{i=0}^{N-1} l_i |l_i\rangle \langle l_i| \tag{2}$$

with $l_0 \leq l_1 \leq ... \leq l_{N-1}$. In particular, by considering the overlap matrix $C_{ij} = |\langle l_i|a_j\rangle|$ we study both how many Laplacian modes are used as models' eigendirections as well as how relevant they are as fraction of the non-trivial spectrum of $\tilde{A}$ (by non-trivial we mean $i \neq 0$ as by construction $|\langle l_0|a_0\rangle| = 1$ for properties of $L$ and $\tilde{A}$). This fraction is quantified by considering $\eta = \frac{\sum_{i \in \mathcal{U} \setminus 0} |a_i|}{\sum_{i=1}^{i=N-1} |a_i|}$ where $\mathcal{U} = \{j| \max_j C_{ij} \geq 0.9 \text{ for } i \in (0, 1, 2, ..., N-1)\}$ with 0.9 being a chosen arbitrary threshold for similarity. Based on these quantities, we define a metric that factors everything together as:

$$\zeta = \eta \sum_{i=1}^{N-1} \Theta \left( \max_j C_{i,j} - 0.9 \right) \tag{3}$$

**Table 1.** Global results obtained from the ADMET group of TDC are presented. Each row corresponds to a specific task, along with the metric used for evaluation. Columns represent different pretrainings considered. Highlighted values denote the best performance achieved among our models, based on the average value as per ranking criterion form the TDC leaderboard. Additionally, cases where our results surpass the top-performing model in the TDC leaderboard are marked with an asterisk (*).

| TDC ADMET Task | METRIC | SCRATCH | ALL | FUKUI_E | NMR | FUKUI_N | CHARGE | HLGAP | MASKED |
|---|---|---|---|---|---|---|---|---|---|
| CACO2_WANG | MAE ↓ | 0.48 ± 0.06 | 0.41 ± 0.03 | 0.45 ± 0.07 | 0.48 ± 0.06 | **0.39 ± 0.02** | 0.40 ± 0.08 | 0.53 ± 0.02 | 0.45 ± 0.01 |
| HIA_HOU | ROC-AUC ↑ | 0.96 ± 0.03 | 0.94 ± 0.05 | 0.93 ± 0.03 | 0.97 ± 0.02 | 0.94 ± 0.02 | 0.95 ± 0.02 | 0.96 ± 0.02 | **0.98 ± 0.01** |
| PGP_BROCCATELLI | ROC-AUC ↑ | 0.87 ± 0.04 | 0.89 ± 0.02 | 0.89 ± 0.03 | 0.86 ± 0.03 | **0.90 ± 0.01** | 0.88 ± 0.02 | 0.86 ± 0.01 | 0.89 ± 0.01 |
| BIOAVAILABILITY_MA | ROC-AUC ↑ | 0.52 ± 0.01 | 0.64 ± 0.05 | 0.64 ± 0.02 | 0.66 ± 0.01 | **0.69 ± 0.05** | 0.62 ± 0.07 | 0.55 ± 0.03 | 0.66 ± 0.05 |
| LIPOPHILICITY_ASTRAZENECA | MAE ↓ | 0.58 ± 0.02 | **0.42 ± 0.01*** | 0.49 ± 0.02 | 0.46 ± 0.01* | 0.48 ± 0.01 | 0.43 ± 0.01* | 0.57 ± 0.02 | 0.47 ± 0.01 |
| SOLUBILITY_AQSOLDB | MAE ↓ | 0.89 ± 0.04 | 0.75 ± 0.01* | 0.80 ± 0.02 | 0.78 ± 0.02 | 0.78 ± 0.02 | **0.75 ± 0.01*** | 0.89 ± 0.02 | 0.76 ± 0.02* |
| BBB_MARTINS | ROC-AUC ↑ | 0.83 ± 0.01 | **0.88 ± 0.02** | 0.86 ± 0.03 | 0.86 ± 0.02 | 0.85 ± 0.02 | 0.87 ± 0.01 | 0.82 ± 0.03 | 0.85 ± 0.02 |
| PPBR_AZ | MAE ↓ | 8.38 ± 0.24 | 7.79 ± 0.24 | 7.92 ± 0.12 | 8.02 ± 0.40 | 7.79 ± 0.28 | **7.57 ± 0.32** | 8.22 ± 0.23 | 8.09 ± 0.13 |
| VDSS_LOMBARDO | SPEARMAN ↑ | 0.58 ± 0.04 | 0.59 ± 0.03 | **0.64 ± 0.02** | 0.61 ± 0.04 | 0.61 ± 0.01 | 0.63 ± 0.03 | 0.59 ± 0.04 | 0.63 ± 0.01 |
| CYP2D6_VEITH | PR-AUC ↑ | 0.43 ± 0.03 | **0.61 ± 0.02** | 0.55 ± 0.03 | 0.56 ± 0.04 | 0.56 ± 0.02 | 0.58 ± 0.04 | 0.47 ± 0.02 | 0.59 ± 0.02 |
| CYP3A4_VEITH | PR-AUC ↑ | 0.73 ± 0.02 | **0.80 ± 0.03** | 0.77 ± 0.03 | 0.78 ± 0.01 | 0.79 ± 0.03 | 0.76 ± 0.04 | 0.74 ± 0.03 | 0.77 ± 0.02 |
| CYP2C9_VEITH | PR-AUC ↑ | 0.63 ± 0.02 | **0.69 ± 0.02** | 0.67 ± 0.02 | **0.69 ± 0.04** | **0.69 ± 0.01** | **0.69 ± 0.02** | 0.66 ± 0.03 | **0.69 ± 0.03** |
| CYP2D6_SUBSTRATE_CARBONMANGELS | PR-AUC ↑ | 0.52 ± 0.01 | 0.58 ± 0.03 | 0.53 ± 0.06 | 0.64 ± 0.06 | 0.57 ± 0.04 | 0.63 ± 0.03 | 0.54 ± 0.04 | **0.66 ± 0.03** |
| CYP3A4_SUBSTRATE_CARBONMANGELS | ROC-AUC ↑ | 0.63 ± 0.07 | 0.64 ± 0.02 | **0.66 ± 0.03** | 0.62 ± 0.02 | 0.61 ± 0.02 | 0.63 ± 0.02 | 0.64 ± 0.03 | 0.65 ± 0.01 |
| CYP2C9_SUBSTRATE_CARBONMANGELS | PR-AUC ↑ | 0.35 ± 0.02 | **0.37 ± 0.04** | 0.32 ± 0.04 | 0.34 ± 0.03 | **0.37 ± 0.04** | 0.36 ± 0.04 | 0.33 ± 0.03 | 0.31 ± 0.03 |
| HALF_LIFE_OBACH | SPEARMAN ↑ | 0.39 ± 0.07 | **0.48 ± 0.06** | 0.48 ± 0.04 | 0.42 ± 0.10 | 0.48 ± 0.03 | 0.47 ± 0.04 | 0.34 ± 0.07 | 0.47 ± 0.06 |
| CLEARANCE_MICROSOME_AZ | SPEARMAN ↑ | 0.49 ± 0.03 | **0.60 ± 0.01** | 0.47 ± 0.06 | 0.57 ± 0.01 | 0.58 ± 0.02 | 0.58 ± 0.01 | 0.46 ± 0.03 | 0.59 ± 0.01 |
| CLEARANCE_HEPATOCYTE_AZ | SPEARMAN ↑ | 0.34 ± 0.04 | 0.46 ± 0.03 | 0.42 ± 0.02 | 0.44 ± 0.04 | 0.41 ± 0.02 | 0.46 ± 0.04 | 0.31 ± 0.02 | **0.47 ± 0.02** |
| HERG | ROC-AUC ↑ | 0.78 ± 0.01 | 0.77 ± 0.06 | 0.73 ± 0.06 | 0.77 ± 0.05 | 0.77 ± 0.02 | 0.79 ± 0.03 | 0.76 ± 0.04 | **0.81 ± 0.04** |
| AMES | ROC-AUC ↑ | 0.72 ± 0.02 | **0.80 ± 0.06** | 0.78 ± 0.02 | **0.80 ± 0.02** | 0.76 ± 0.01 | **0.80 ± 0.01** | 0.73 ± 0.01 | 0.79 ± 0.01 |
| DILI | ROC-AUC ↑ | 0.86 ± 0.02 | 0.88 ± 0.03 | 0.86 ± 0.04 | 0.89 ± 0.03 | 0.82 ± 0.04 | 0.85 ± 0.04 | 0.87 ± 0.01 | **0.91 ± 0.01** |
| LD50_ZHU | MAE ↓ | 0.61 ± 0.02 | 0.57 ± 0.02 | 0.60 ± 0.01 | **0.56 ± 0.02** | 0.60 ± 0.02 | 0.57 ± 0.01 | 0.60 ± 0.03 | 0.57 ± 0.02 |

where $\Theta$ is the Heaviside function. We then evaluate $\zeta$ averaged over the test set of each downstream task reporting per each architecture the distribution across tasks for fixed pretraining condition, and also analyse for every task if the model ranking in peformance correlates with the ranking coming from the evaluation of the average $\zeta$ over that test set. Finally, for this reason we make use of the Spearman's rank coefficient and consider performances as the higher the better (e.g. we consider MAE with a negative sign but ROC-AUC with positive sign).

# 3    Results and Discussion

**Pretraining on Atom-Resolved Tasks Give the Best Overall Performances.** Model performances obtained for the downstream tasks are summarized in Table 1. The table reveals, among other things, that the models trained from scratch or pretrained on the HOMO-LUMO gap (molecular proeperty) are never among the top performers. The superior performance of the models pretrained on atom-level properties is remarkable considering that the HOMO-LUMO gap dataset contains $\sim$ 20 times more molecules than present in the dataset used for pretraining on atomic QM properties and for masking.

**The Right Atomic QM Pretraining Usually Gives the Best Performances.** In the same table it is also possible to count which pretraining gives most frequently the best results. Despite the fact that results can be quite close,

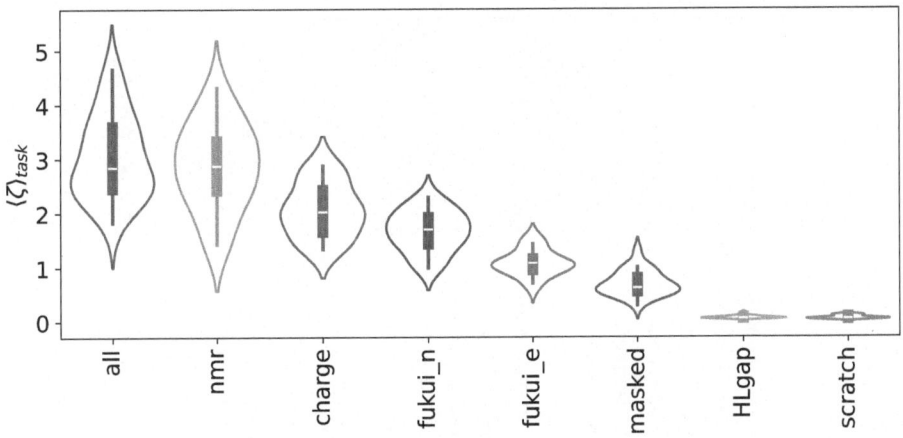

**Fig. 2.** Violin plots of the average value of $\zeta$ on each task in the ADMET group of TDC ($\langle\zeta_{task}\rangle$) for every Graphormer model considered.

if we rank the models by average value of the metric as done in the TDC leaderboard, the models that demonstrate the highest number of top performances were pretrained using all the atomic QM properties with 10 and pretrained with masking with 6 top results, respectively. Atom-level QM pretraining as a group reveals even higher superiority over studied alternatives: that is in 17 out of 22 downstream tasks the correct choice of atom-level QM pretraining provides the top performant model.

**Atom-Level QM Pretraining Boosts the Spectral Perception of Molecules.** We evaluate the metric $\zeta$ defined in Eq. 3 as described in the Sect. 2 obtaining a distribution of 22 values over the downstream tasks per each pretraining. The result is reported in Fig. 2 as a set of violin plots. Firstly, we clearly see that models trained from scratch or pretrained on HOMO-LUMO gap present values of $\zeta$ that are close to 0 indicating little to no presence of non-trivial Laplacian eigenmodes in the spectrum of their $\tilde{A}$ matrix. On the contrary, every atom-resolved pretrained model (including masking) presents nonzero values of $\zeta$ across the downstream tasks raging from $\sim 1$ to $\sim 5$. Within these last group of models we can clearly notice how pretraining on the atom-level QM properties provides the strongest boost in perception of graph Laplacian eigenmodes. In particular, the model pretrained using all properties in a multi-task fashion and using only NMR data present the highest values of $\zeta$, followed by the models pretrained on charges, nucleophilic and electrophilic Fukui functions.

**A Better Spectral Perception of Molecules Usually Correlates with Better Performances.** As described in the Sect. 2, we proceed to analyse the Spearman's rank coefficient $r_S$ between $\zeta$ and performances in each task using the 8 datapoints coming from the different pretraining methods. The results

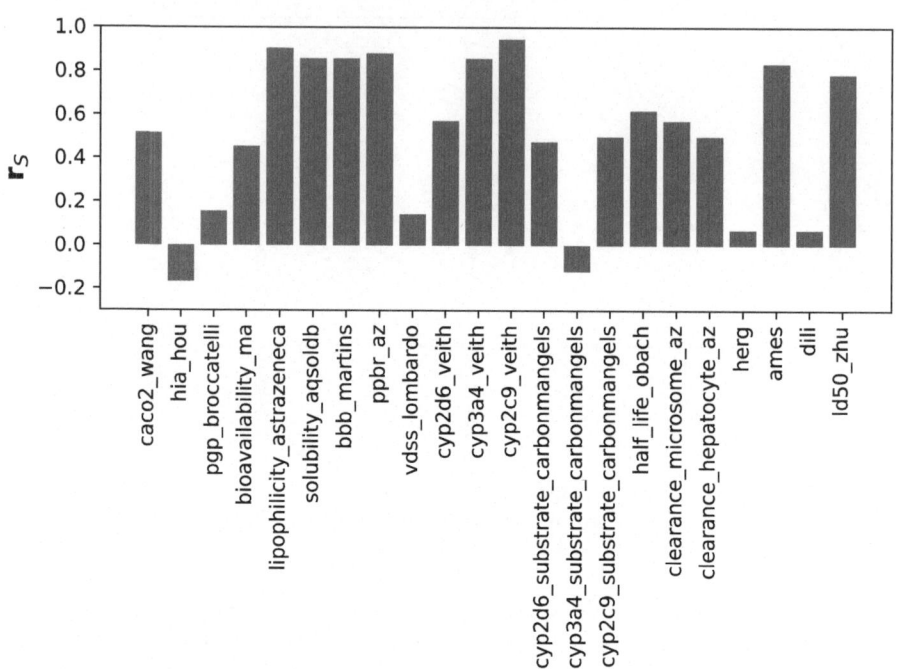

**Fig. 3.** Spearman's rank coefficient $r_S$ between the average $\zeta$ value per each model and the correspondent performance per each task in the ADMET group of TDC.

are reported in Fig. 3. We can see that for most tasks (20 out of 22) the value of $r_S$ is positive, with 13 tasks presenting $r_S \geq 0.5$ and 8 tasks presenting $r_S \geq 0.75$. These results are a strong indication that models with a better spectral perception of the molecular graph also demonstrate better performances across different tasks.

## 4    Conclusions

A Graphormer neural network was pretrained on several tasks to improve its performance in modelling molecular ADMET properties that are relevant to drug discovery using the TDC dataset containing 22 downstream tasks. It was found that out of studied methods, pretraining on atom-level QM properties such as atomic charges, NMR shielding constants and Fukui indexes, or using a masking task similar to the one used in BERT model, significantly improve the performance in comparison to the non-pretrained model. One of atom-level QM property pretraining tasks was found to yield the best results for 17 out of 22 downstream tasks. For comparison, pretraining on a much larger dataset of calculated HOMO-LUMO gaps, a molecular electronic property, brings little or no improvement. Finally, through a spectral analysis of the Attention Rollout matrices, we showed how pretraining on atom-level QM properties improves the model perception of spectral properties of the input molecular graph. In particular, by defining an appropriate metric, we show that this effect correlates with the model performance on most of the downstream tasks.

**Acknowledgments.** This study was funded under the European Union's Horizon 2020 research and innovation program under the Marie Skłodowska-Curie grant agreement No 956832, "Advanced Machine learning for Innovative Drug Discovery" (AIDD).

**Disclosure of Interests.** The authors have no competing interests to declare that are relevant to the content of this article.

# References

1. Abnar, S., Zuidema, W.H.: Quantifying attention flow in transformers (2020). https://arxiv.org/abs/2005.00928
2. Beck, M.E.: Do fukui function maxima relate to sites of metabolism? A critical case study. J. Chem. Inform. Model. **45**(2), 273–282 (2005). https://doi.org/10.1021/ci049687n, pMID: 15807488
3. Born, J., et al.: Chemical representation learning for toxicity prediction. Digit. Disc. **2**, 674–691 (2023). https://doi.org/10.1039/D2DD00099G
4. Broccatelli, F., Trager, R., Reutlinger, M., Karypis, G., Li, M.: Benchmarking accuracy and generalizability of four graph neural networks using large in vitro ADME datasets from different chemical spaces. Mol. Inform. **41**(8), 2100321 (2022). https://doi.org/10.1002/minf.202100321, https://onlinelibrary.wiley.com/doi/abs/10.1002/minf.202100321
5. Bule, M., Jalalimanesh, N., Bayrami, Z., Baeeri, M., Abdollahi, M.: The rise of deep learning and transformations in bioactivity prediction power of molecular modeling tools. Chem. Biol. Drug Des. **98**(5), 954–967 (2021). https://doi.org/10.1111/cbdd.13750, https://onlinelibrary.wiley.com/doi/abs/10.1111/cbdd.13750
6. Chen, H., Engkvist, O., Wang, Y., Olivecrona, M., Blaschke, T.: The rise of deep learning in drug discovery. Drug Discov. Today **23**(6), 1241–1250 (2018). https://doi.org/10.1016/j.drudis.2018.01.039, https://www.sciencedirect.com/science/article/pii/S1359644617303598
7. Chuang, K.V., Gunsalus, L.M., Keiser, M.J.: Learning molecular representations for medicinal chemistry. J. Med. Chem. **63**(16), 8705–8722 (2020). https://doi.org/10.1021/acs.jmedchem.0c00385, pMID: 32366098
8. David Z Huang, J.C.B., Bahmanyar, S.S.: The challenges of generalizability in artificial intelligence for ADME/TOX endpoint and activity prediction. Expert Opin. Drug Discov. **16**(9), 1045–1056 (2021). https://doi.org/10.1080/17460441.2021.1901685, pMID: 33739897
9. Devlin, J., Chang, M.W., Lee, K., Toutanova, K.: BERT: Pre-training of Deep Bidirectional Transformers for Language Understanding (2018)
10. Ektefaie, Y., Shen, A., Bykova, D., Marin, M., Zitnik, M., Farhat, M.: Evaluating generalizability of artificial intelligence models for molecular datasets. bioRxiv (2024). https://doi.org/10.1101/2024.02.25.581982, https://www.biorxiv.org/content/early/2024/02/28/2024.02.25.581982
11. Fabian, B., et al.: Molecular representation learning with language models and domain-relevant auxiliary tasks. In: Proceedings of the NeurIPS 2020 Workshop on Machine Learning for Molecules (2020)

12. Glavatskíkh, M., Leguy, J., Hunault, G., Cauchy, T., Da Mota, B.: Dataset's chemical diversity limits the generalizability of machine learning predictions. J. Cheminform. **11**(1), 69 (2019). https://doi.org/10.1186/s13321-019-0391-2

13. Guan, Y., et al.: Regio-selectivity prediction with a machine-learned reaction representation and on-the-fly quantum mechanical descriptors. Chem. Sci. **12**(6), 2198–2208 (2021). https://doi.org/10.1039/d0sc04823b

14. Hoja, J., et al.: Qm7-x, a comprehensive dataset of quantum-mechanical properties spanning the chemical space of small organic molecules. Sci. Data **8**(1), 43 (2021). https://doi.org/10.1038/s41597-021-00812-2

15. Hu, W., et al.: Strategies for pre-training graph neural networks. In: International Conference on Learning Representations (2020). https://openreview.net/forum?id=HJlWWJSFDH

16. Huang, K., et al.: Artificial intelligence foundation for therapeutic science. Nat. Chem. Biol. **18**(10), 1033–1036 (2022). https://doi.org/10.1038/s41589-022-01131-2

17. Isert, C., Atz, K., Jiménez-Luna, J., Schneider, G.: QMugs, quantum mechanical properties of drug-like molecules. Sci. Data **9**(1) (2022). https://doi.org/10.1038/s41597-022-01390-7

18. Jayatunga, M.K., Xie, W., Ruder, L., Schulze, U., Meier, C.: Ai in small-molecule drug discovery: a coming wave. Nat. Rev. Drug Discov. **21**, 175–176 (2022)

19. Kaufman, B., et al.: COATI: multimodal contrastive pretraining for representing and traversing chemical space. J. Chem. Inform. Model. **64**(4), 1145–1157 (2024). https://doi.org/10.1021/acs.jcim.3c01753, pMID: 38316665

20. Li, M.M., Huang, K., Zitnik, M.: Graph representation learning in biomedicine and healthcare. Nat. Biomed. Eng. **6**(12), 1353–1369 (2022). https://doi.org/10.1038/s41551-022-00942-x

21. Medrano Sandonas, L., et al.: Dataset for quantum-mechanical exploration of conformers and solvent effects in large drug-like molecules. Sci. Data **11**(1), 742 (2024)

22. Müller, L., Galkin, M., Morris, C., Rampášek, L.: Attending to graph transformers. Transactions on Machine Learning Research (2024). https://openreview.net/forum?id=HhbqHBBrfZ

23. Nakata, M., Shimazaki, T.: PubChemQC project: a large-scale first-principles electronic structure database for data-driven chemistry. J. Chem. Inf. Model. **57**(6), 1300–1308 (2017). https://doi.org/10.1021/acs.jcim.7b00083

24. Nugmanov, R., Dyubankova, N., Gedich, A., Wegner, J.K.: Bidirectional graphormer for reactivity understanding: neural network trained to reaction atom-to-atom mapping task. J. Chem. Inform. Model. **62**(14), 3307–3315 (2022). https://doi.org/10.1021/acs.jcim.2c00344, pMID: 35792579

25. Wang, Y., Xu, C., Li, Z., Barati Farimani, A.: Denoise pretraining on nonequilibrium molecules for accurate and transferable neural potentials. J. Chem. Theory Comput. **19**(15), 5077–5087 (2023). https://doi.org/10.1021/acs.jctc.3c00289, pMID: 37390120

26. Xia, J., et al.: Mole-BERT: rethinking pre-training graph neural networks for molecules. In: The Eleventh International Conference on Learning Representations (2023). https://openreview.net/forum?id=jevY-DtiZTR

27. Xia, J., Zhu, Y., Du, Y., Li, S.Z.: A systematic survey of chemical pre-trained models. In: Elkind, E. (ed.) Proceedings of the Thirty-Second International Joint Conference on Artificial Intelligence, IJCAI-23. pp. 6787–6795. International Joint Conferences on Artificial Intelligence Organization (2023). https://doi.org/10.24963/ijcai.2023/760, survey Track
28. Ying, C., et al.: Do transformers really perform badly for graph representation? In: Advances in Neural Information Processing Systems, vol. 34, pp. 28877–28888. Curran Associates, Inc. (2021). https://proceedings.neurips.cc/paper_files/paper/2021/file/f1c1592588411002af340cbaedd6fc33-Paper.pdf

**Open Access** This chapter is licensed under the terms of the Creative Commons Attribution 4.0 International License (http://creativecommons.org/licenses/by/4.0/), which permits use, sharing, adaptation, distribution and reproduction in any medium or format, as long as you give appropriate credit to the original author(s) and the source, provide a link to the Creative Commons license and indicate if changes were made.

The images or other third party material in this chapter are included in the chapter's Creative Commons license, unless indicated otherwise in a credit line to the material. If material is not included in the chapter's Creative Commons license and your intended use is not permitted by statutory regulation or exceeds the permitted use, you will need to obtain permission directly from the copyright holder.

# Balancing Imbalanced Toxicity Models: Using MolBERT with Focal Loss

Muhammad Arslan Masood[1,2(✉)] [iD], Samuel Kaski[2,3] [iD], Hugo Ceulemans[1] [iD],
Dorota Herman[1] [iD], and Markus Heinonen[2] [iD]

[1] Drug Discovery Data Sciences, Janssen Pharmaceutica NV, Turnhoutseweg 30,
2340 Beerse, Belgium
[2] Department of Computer Science, Aalto University, Espoo, Finland
arslan.masood@aalto.fi
[3] Department of Computer Science, University of Manchester, Manchester, UK
https://www.aalto.fi/en/department-of-computer-science

**Abstract.** Drug-induced liver injury (DILI) presents a multifaceted
challenge, influenced by interconnected biological mechanisms. Current
DILI datasets are characterized by small sizes and high imbalance, pos-
ing difficulties in learning robust representations and accurate modeling.
To address these challenges, we trained a multi-modal multi-task model
integrating preclinical histopathologies, biochemistry (blood markers),
and clinical DILI-related adverse drug reactions (ADRs). Leveraging pretrained BERT models, we extracted representations covering a broad
chemical space, facilitating robust learning in both frozen and fine-
tuned settings. To address imbalanced data, we explored weighted Binary
Cross-Entropy (w-BCE) and weighted Focal Loss (w-FL) . Our results
demonstrate that the frozen BERT model consistently enhances performance across all metrics and modalities with weighted loss functions
compared to their non-weighted counterparts. However, the efficacy of
fine-tuning BERT varies across modalities, yielding inconclusive results.
In summary, the incorporation of BERT features with weighted loss functions demonstrates advantages, while the efficacy of fine-tuning remains
uncertain.

**Keywords:** Toxicity · DILI · BERT · Focal loss

## 1  Introduction and Background

Thalidomide, the tragedy of birth defects led the foundation of systematic testing
of drugs safety prior to marketing (Kim and Scialli, 2011). Pharmacovigilance
efforts start with in-vitro and in-vivo studies during the drug development stage,
continue through clinical trial and post-marketing surveillance.

**Supplementary Information** The online version contains supplementary material
available at https://doi.org/10.1007/978-3-031-72381-0_8.

© The Author(s) 2025
D.-A. Clevert et al. (Eds.): AIDD 2024, LNCS 14894, pp. 82–97, 2025.
https://doi.org/10.1007/978-3-031-72381-0_8

The liver, as the primary organ affected by xenobiotics, plays a crucial role in drug metabolism (Stanley, 2017). Drug-induced liver injury (DILI) stands as a significant cause of late-stage drug failure and post-marketing drug withdrawal (Watkins, 2011; Parasrampuria et al., 2018). Toxic compounds can be categorized into intrinsic toxins, whose toxicity can be modeled based on chemical information, and idiosyncratic toxins, which pose challenges in both preclinical and clinical modeling due to their unpredictable effects influenced by genetic variations (Lancaster et al., 2015; Parasrampuria et al., 2018). Over the years, several methods have been developed to model DILI using molecular structure and various fingerprints (Cruz-Monteagudo et al., 2008; Chen et al., 2013b; Xu et al., 2015; Ai et al., 2018; Wang et al., 2019; Asilar et al., 2020). Combining other modalities with molecular features, such as transcriptomics (Wang et al., 2019a), physicochemical properties (Ekins et al., 2010; Chen et al., 2013a), and selected in-vitro assays (Williams et al., 2020), has been shown to provide robust DILI models. During the drug design process, toxicity assessment spans multiple stages, encompassing in-vitro assays, preclinical investigations, and clinical trials. Toxicity presents across diverse endpoints and species, thus prompting a multitask approach for data integration and cross-modality learning. This strategy has demonstrated promise in extracting toxicity patterns by jointly considering various dose administration methods, endpoints, and species, particularly in acute toxicity modeling (Sosnin et al., 2019; Jain et al., 2021). Moreover, extending this approach to incorporate joint learning from in-vitro, in-vivo, and clinical data has improved balanced accuracy (as defined in Eq. 7) of the ClinTox dataset. (Sharma et al., 2023).

Class imbalance is a prevalent issue in toxicity datasets, where negative instances vastly outnumber positive ones. This disparity makes machine learning models inaccurate, as classifiers trained on imbalanced data tend to prioritize the majority class, leading to ineffective performance on the minority class (Rawat and Mishra, 2022). To address this, various strategies are employed, including resampling techniques like oversampling and undersampling. Oversampling methods such as Synthetic Minority Oversampling Technique (SMOTE) artificially increase the number of minority instances (Chawla et al., 2002), while undersampling involves reducing the number of majority instances (Laveti et al., 2021; Lee and Seo, 2022). However, both approaches have drawbacks; undersampling may lead to loss of valuable data, while oversampling can be computationally intensive (Rawat and Mishra, 2022). Cost-Sensitive Learning (CSL) can also be used as this method assigns higher costs to samples from the minority class (Elkan, 2001; López et al., 2012). Unlike resampling techniques, CSL maintains the original data distribution while enhancing computational efficiency. CSL, coupled with traditional machine learning algorithms such as Random Forest (RF) and Support Vector Machine (SVM), has been used for drug discovery application, including compound activity estimation (Alashwal and Lucman, 2020), CYP450 modeling (Eitrich et al., 2007), and Drug-Induced Liver Injury (DILI) modeling (Moein et al., 2023), demonstrating improvements in some cases. In the realm of deep learning, binary-cross-entropy loss serves as a

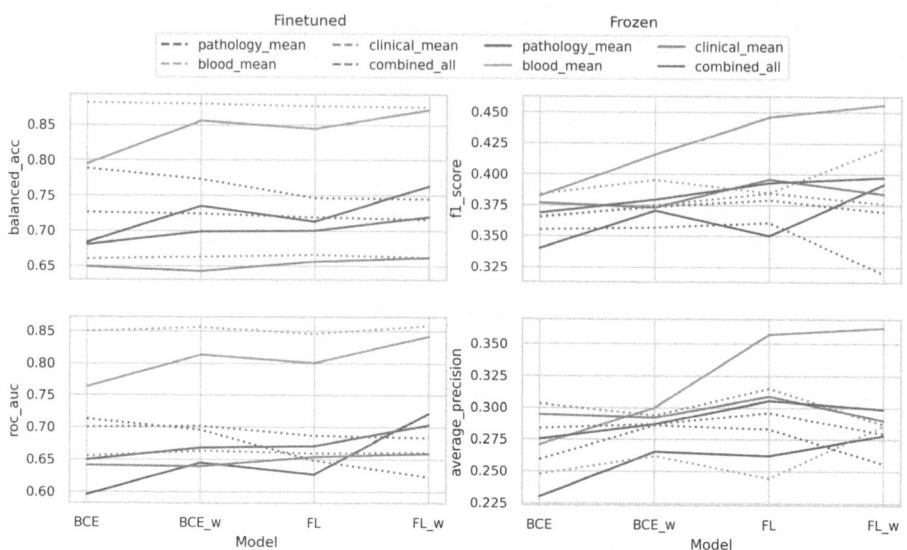

**Fig. 1.** Mean performance across all tasks was evaluated using multiple metrics. Solid lines represent Frozen-BERT, while dotted lines indicate fine-tuned BERT. Performance improved from BCE to weighted-BCE, Focal loss, and weighted Focal loss for the Frozen-BERT model. However, this trend is inconsistent for fine-tuned BERT. In certain tasks, such as pathological ones, the fine-tuned model outperformed Frozen-BERT. In other cases, negative transfer is observed.

common choice for training binary classification models, often augmented with a weighting factor to elevate the cost of positive instances, thus ensuring a balanced contribution from both classes in the overall loss function. Focal loss represents a refinement of BCE loss, introducing a modulating factor that aids in distinguishing between easy and difficult examples naturally favours minority class Lin et al. (2018).

One method for representing three-dimensional chemical structures as text strings is the Simplified Molecular Input Line Entry System (SMILES), which employs a defined set of ordered rules and specific syntax Weininger (1988). The chemical characteristics of a compound $\mathbf{x}_c$ can be described through various modalities. Encoding schemes like Mold, PaDel, RDF, ECFC, and Marvin molecular descriptors have been developed to capture molecular structural properties. Despite their individual successes, there's no universal encoding scheme or algorithm to *rule them all* (Gao et al., 2020). In drug development, small-scale datasets often fail to adequately represent the vast chemical space, leading to models trained on handcrafted features that struggle to generalize to unseen chemical spaces (Moein et al., 2023). To address this limitation, researchers leverage representations derived from large amounts of unlabeled data (Harnik and Milo, 2024). Various models such as Variational Autoencoders (VAEs) (Kingma and Welling, 2013), Normalizing Flows (NFs) (Rezende and Mohamed, 2015), and Generative Adversarial Networks (GANs) (Goodfellow et al., 2014) aim to uncover low-dimensional latent representations $\phi(\mathbf{x}) \in \mathbb{R}^d$ of complex, high-dimensional objects $\mathbf{x} \in \mathbb{R}^D$, where $d \ll D$ (Ruthotto and Haber, 2021). These models transform data into a vectorized space, generating concise and well-structured representations that encompass broader chemical space (Li et al., 2022). To facilitate learning of underlying chemistry, various pretext tasks are carefully designed, including input translation between modalities (Winter et al., 2019; Yang et al., 2019), input reconstruction (Wang et al., 2019b; Li and Fourches, 2020; Maziarka et al., 2020), and recovering masked or corrupted input (Liu et al., 2023).

In recent years, various transformer-based models have been applied to molecular representation learning (Chithrananda et al., 2020; Li and Jiang, 2021; Ahmad et al., 2022; Irwin et al., 2022) with many studies opted for transformer based BERT architecture (Li and Jiang, 2021; Liu et al., 2023; Shermukhamedov et al., 2023). BERT (Bidirectional Encoder Representations from Transformers) is pre-trained on large text corpora using two objectives, defined by Devlin et al. (2019) as the "masked language model" (MLM) and "next sentence prediction" (NSP) task. During pre-training, BERT learns bidirectional contextual embeddings for each token, capturing nuanced word meanings within the sentence context. Utilizing the Transformer architecture, BERT employs self-attention mechanisms to dynamically weigh word importance. By fine-tuning on task-specific labeled data, BERT adapts its learned representations to various downstream natural language processing tasks, achieving state-of-the-art performance. BERT can learn molecular representations by treating molecular structures as token sequences. Pre-training BERT on large molecular datasets with appropriate objectives, such as incorporating physicochemical properties or molecule relationships, enables it to learn robust chemical representations (Fabian et al., 2020). Fine-tuning the pre-trained BERT model on small task-specific labeled data can provide improved performance in some drug discovery applications (Liu et al., 2023).

## 2 Materials and Method

### 2.1 Datasets

The preclinical study integrates liver histopathology endpoints from the TG-Gates dataset (Igarashi et al., 2015), covering 170 compounds administered to rats across varying concentrations and exposure conditions, later expanded to 430 compounds with re-annotated INHAND labels (Moein et al., 2023). Out of 55 liver endpoints, we focus on 12 for this study. We extend preclinical tasks by incorporating selected blood markers (ALP, AST, ALT, GTP, TC, TG, TBIL, DBIL) from biochemistry database provided by TG-GATES converting both histopathological and bloodmarker labels into binary labels using expert-derived thresholds. Additionally, we enrich preclinical data with DILI related adverse drug reactions (ADRs) extracted from the SIDER dataset , comprising 6060 ADRs associated with 1430 drugs (Kuhn et al., 2016). Further details regarding

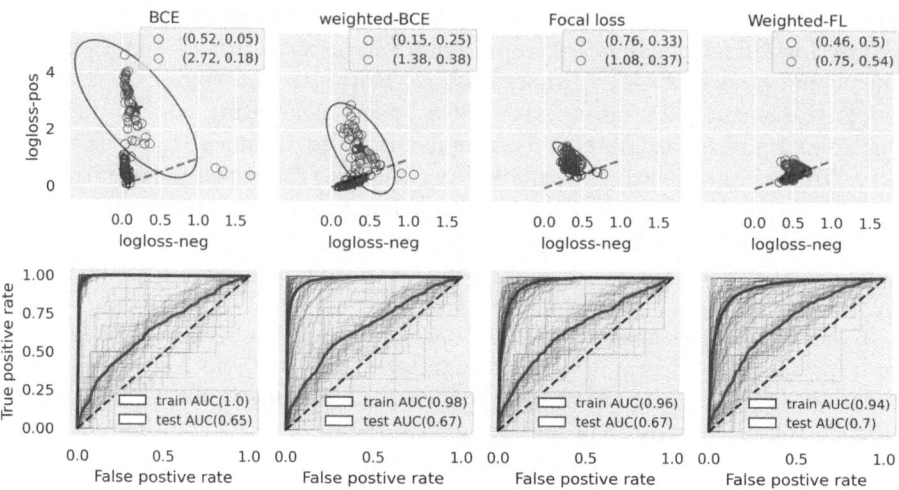

**Fig. 2.** Results from Frozen-BERT model. **Top row**: The log-loss analysis of positive and negative data points is depicted in this plot, where blue and green colors represent the training and testing data, respectively. Task-wise means are represented by small blobs, while ellipses indicate the 95% confidence interval. This visualization revealed that the network tends to be biased towards the majority class (negatives), leading to significantly lower log-loss for negatives, particularly evident with Binary Cross-Entropy (BCE). Transitioning to weighted Binary Cross-Entropy (BCE-W), the model is forced to equally prioritize both negatives and positives, resulting in a decrease in log-loss for positives. Focal Loss naturally emphasizes on hard examples, which, in this context, are positive examples. Weighted Focal Loss further supported the model by applying additional weighting to positive examples, as a result further reduced in logloss of positive instances. **Bottom row**: This plot presents the ROC-AUC for the train and test sets. Light lines represent task-wise ROC-AUC, while the thick line represents the mean ROC-AUC across all tasks. Weighted Focal Loss provided the highest validation ROC-AUC

task selection, binarization, and distributions are available in the supplementary material.

## 2.2   Loss Functions

We consider a modeling problem from molecules $\mathbf{x}$ to binary toxicity profiles $\mathbf{y} \in \{0,1\}^P$ of $P = 50$ endpoints from a dataset $D = \{(\mathbf{x}_n, \mathbf{y}_n)\}_{n=1}^N$ of size $N \approx 2000$. We assume a function $f(\mathbf{x}; \theta) \in [0,1]^P$ that outputs separate probabilities for endpoints, and we use a shorthand $f_{np} = f(\mathbf{x}_n; \theta)_p$.

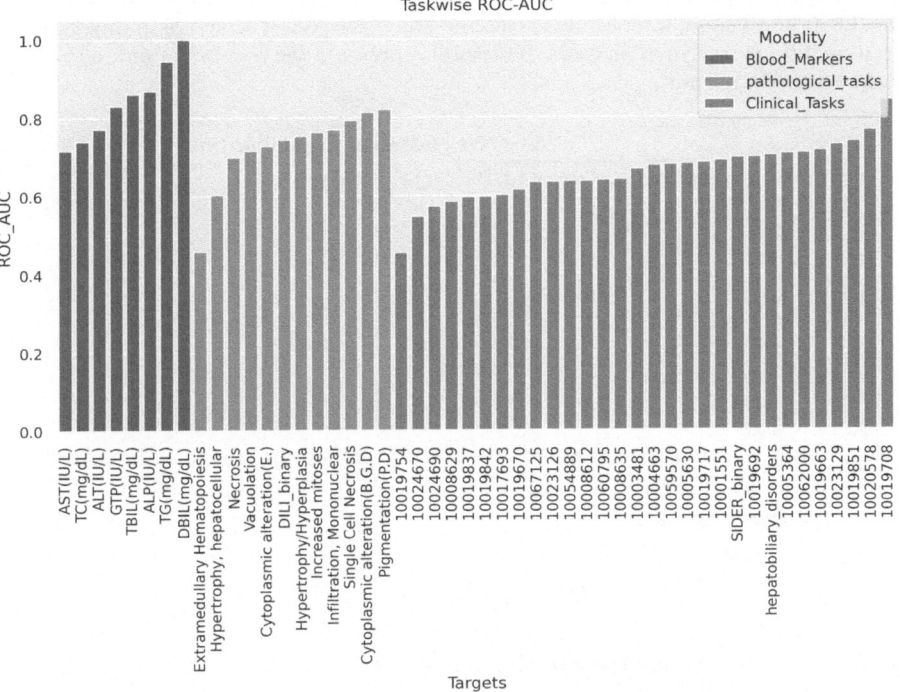

**Fig. 3.** This plot shows task-wise performance, with colors representing different modalities and tasks organized from lowest to highest ROC score. This plot also shows that clinical tasks were the most challenging to model. Additionally, the model failed to learn two tasks: Extramedullary (pathological) and 100197554 (clinical)

**Binary-Cross-Entropy Loss (BCE).** The binary cross-entropy (BCE) training loss is appropriate for this problem

$$\mathcal{L}_{\text{BCE}} = \sum_{p=1}^{P} \sum_{n=1}^{N} y_{np} \log \sigma(f_{np}) + (1 - y_{np}) \log(1 - \sigma(f_{np})) \qquad (1)$$

**Weighted-BCE.** The toxicity datasets are generally zero-inflated with negatives being much more common, however, the BCE treats each observation as equally important, and will lead the model to focus more on negatives. We can tackle this positive-negative imbalance by overweighting the positive datapoints within each endpoint,

$$\mathcal{L}_{\text{BCE}}^{\text{w}} = \sum_{n=1}^{N} \sum_{p=1}^{P} w_p^{+} y_{np} \log \sigma(f_{np}) + (1 - y_{np}) \log(1 - \sigma(f_{np})) \qquad (2)$$

where $w_p^{+} = N_{p-}/N_{p+} \in \mathbb{R}^{+}$ is the inverse ratio of positives $N_{p+}$ to negatives $N_{p-}$ in endpoint $p$. Here we only upscale the positives while leaving negatives

**Table 1.** The distribution of positive and negative samples across each modality. For visualization purpose, a molecule is classified as positive if it is active in any task, and negative if it is inactive in all tasks. The total represents the number of unique SMILES in the complete dataset.

|  | positive | negative | Modality Sum |
|---|---|---|---|
| Pathologies | 176 | 234 | 410 |
| Biochemistry (blood markers) | 124 | 286 | 410 |
| Clinical | 749 | 470 | 1219 |
| Total | 1049 | 990 | 1554* |

intact. This loss ensures that within-class positive and negative observations have equal mass. Further, we can try to find the balancing, by selecting optimal $\alpha$ through cross validation

$$w_p^+ = \alpha \frac{N_{p-}}{N_{p+}} + (1 - \alpha)1 \tag{3}$$

where $\alpha \in [0, 1]$ denotes the positive balancing.

**Focal Loss.** In scenarios of significant class imbalance, mere weighting can be insufficient, as it fails to discriminate between easy and challenging examples, thereby risking the overwhelming of gradients by the dominant class. A remedy for this issue is focal loss, initially devised for object detection within images (Lin et al., 2018). This approach incorporates a modulating parameter alongside cross-entropy loss, thereby decrease the influence of accurately classified examples and consequently mitigating their overall impact. This modulating factor can be adopted and integrated into our binary cross-entropy loss framework.

$$\mathcal{L}_{\mathrm{FL}} = \sum_{n=1}^{N} \sum_{p=1}^{P} (1 - \sigma(f_{np}))^\gamma \, y_{np} \log \sigma(f_{np}) + \sigma(f_{np})^\gamma (1 - y_{np}) \log (1 - \sigma(f_{np}))$$

$$\tag{4}$$

**Weighted Focal Loss.** Focal loss can also be assisted by incorporating positive weighting as described earlier.

$$\mathcal{L}_{\mathrm{FL}}^{\mathrm{w}} = \sum_{n=1}^{N} \sum_{p=1}^{P} w_p^+ (1 - \sigma(f_{np}))^\gamma \, y_{np} \log \sigma(f_{np}) + \sigma(f_{np})^\gamma (1 - y_{np}) \log (1 - \sigma(f_{np}))$$

$$\tag{5}$$

## 2.3   Models

**Baseline.** We are using Random Forest as our baseline. Random Forest is a robust baseline as it combines decision trees through ensemble learning, reducing overfitting and providing reliable results. Additionally, Random Forest maintains interpretability and scales efficiently for large datasets. To optimize performance, we conducted individual task-specific hyperparameter searches and presented the mean results across all tasks in Table 2. The hyperparameter search space details are provided in supplementary Table 2.

**MolBERT.** The MolBERT model Fabian et al. (2020), an adaptation of the BERT architecture Devlin et al. (2019), consists of 12 attention heads, 12 layers, and a 768-dimensional hidden layer, containing 85 million parameters. It is primarily optimized for the masked token estimation, employing cross-entropy loss. Additionally, it incorporates physicochemical properties computed via RDKit as an auxiliary task, with optimization achieved through mean squared error. The final loss function is determined by the arithmetic mean of all individual task losses. This model is pretrained for 100 epochs using the Adam optimizer.

**MLP Head.** This MLP head consists of an input-hidden-output layers, where $\mathbf{x}_0$ is initialized as the input features $\mathbf{x}$, which can be either BERT features or ECFP . We utilize dropout for regularization, batch normalization for training stability, and the rectified linear unit (ReLU) activation function as the default activation. Additionally, the network incorporates a skip connection, merging the input and output of the hidden layer, enhancing information flow. Finally, the output layer generates logits, which can be transformed into probabilities by passing through a sigmoidal activation function.

$$\mathbf{x}_0 = \mathbf{x} \quad \text{BERT features or ECFP}$$
$$\mathbf{x}_\ell = \text{Dropout}(\text{ReLU}(\text{BatchNorm}(W_\ell \mathbf{x}_0 + \mathbf{b}_\ell)))$$
$$\tilde{\mathbf{x}}_{\ell+1} = \text{BatchNorm}(W_{\ell+1}\mathbf{x}_\ell + \mathbf{b}_{\ell+1}) \tag{6}$$
$$\mathbf{x}_{\ell+1} = \text{Dropout}(\text{ReLU}(\mathbf{x}_\ell + \tilde{\mathbf{x}}_{\ell+1}))$$
$$x_{out} = W_{\ell+2}\mathbf{x}_{\ell+1} + \mathbf{b}_{\ell+1}$$

The hyper-parameters of this model are given in Table 1 in supplementary.

## 2.4   Feature Extraction

**ECFP Fingerprints.** ECFP or Extended-Connectivity Fingerprints (Rogers and Hahn, 2010), is a method used in cheminformatics to represent molecular structures as binary fingerprints, capturing structural information by encoding the presence or absence of substructural features within a specified radius around each atom. Through iterative traversal of the molecular structure, unique substructural fragments are identified and hashed into a fixed-length bit vector, generating a binary fingerprint where each bit indicates the presence or absence of a specific substructural fragment. We encoded each molecule into fix 1024 dimensional binary vector by using radius 6. We have compared ECFP fingerprints with BERT features explained below.

**BERT Features.** We encoded preclinical and clinical SMILES into continuous features, utilizing a large transformer model MolBERT, pretrained on 1.6 million SMILES via masking, alongside physicochemical properties (Fabian et al., 2020). Extracting a pooled output of dimension 764 from the pretrained model, we employed these features to train an MLP head. This strategy allowed us to leverage a significant volume of unlabeled data,and encapsulated the contextual information of larger chemical space.

## 2.5   Evaluation

Here, we briefly sketch the evaluation metrics used in model selection and to report final results.

*Balanced Accuracy.* Given the imbalance between positive and negative instances, using accuracy as a performance metric becomes inadequate. Therefore, we chose balanced accuracy, which represents the arithmetic mean of sensitivity (true positive rate) and specificity (true negative rate). We compute the balanced accuracy at varying thresholds for each task and select the threshold $(\tau_p^{\max})$ that yields the highest balanced accuracy.

$$\text{BA}(\tau_p) = \frac{1}{2}\left(\text{Sensitivity}(\tau_p) + \text{Specificity}(\tau_p)\right)$$
$$\tau_p^{\max} = \arg\max_{\tau_p} \text{BA}(\tau_p)$$
$$\text{BA} = \frac{1}{P}\sum_{p=1}^{p}\text{BA}(\tau_p^{\max})_p \tag{7}$$

*ROC AUC.* The ROC curve, generated by plotting true positive rates (TPR) against false positive rates (FPR) at various thresholds$(\tau_p)$, illustrates the trade-off in model performance. The area under this curve (ROC AUC) condenses the curve's information into a single value, ranging between 0.5 (no discrimination) and 1.0 (ideal discrimination).

*AUPR.* The ROC-AUC curve can yield overly optimistic results with highly imbalanced datasets, thus we used Precision-Recall (PR) curves (Davis and Goadrich, 2006; Forman and Scholz, 2010). The Average Precision (AP) score provides a summary of a precision-recall curve by calculating the weighted mean of precisions achieved at each threshold, with the increase in recall from the previous threshold used as the weight (Zhu, 2004):

$$\text{AP} = \sum_{n}(R_n - R_{n-1})P_n \tag{8}$$

where $P_n$ and $R_n$ denote the precision and recall at the $n$-th threshold, respectively. We selected the optimal hyperparmeters based on AP-score

*F1-score.* This metric combines the precision and recall using the harmonic mean. To select the optimal threshold, we followed the similar procedure to balance accuracy

$$\text{F1 score} = 2 \times \frac{\text{Precision}(\tau_p) \times \text{Recall}(\tau_p)}{\text{Precision}(\tau_p) + \text{Recall}(\tau_p)} \tag{9}$$

*Log-Loss.* To compute the loss of positive and negative instances for each task, we use the following equations:

$$\mathcal{L}_{\text{pos}}^p = \frac{1}{N_{\text{pos}}} \sum_{n=1}^{N} (y_{np} \log \sigma(f_{np}))$$
$$\mathcal{L}_{\text{neg}}^p = \frac{1}{N_{\text{neg}}} \sum_{n=1}^{N} ((1 - y_{np}) \log(1 - \sigma(f_{np}))) \tag{10}$$

# 3   Results and Discussions

**Table 2.** Comparison of different loss functions with ECFP and BERT features. We also showed the effect of BERT fine-tuning

| Model | Loss type | | | | Features | | Finetuning | Metrics | | | |
|---|---|---|---|---|---|---|---|---|---|---|---|
| | BCE | BCE$^w$ | FL | FL$^w$ | ECFP | BERT | | BA | F1 | ROC | AP |
| RF | - | - | - | - | - | - | - | 0.67 ± 0.002 | 0.36 ± 0.003 | 0.65 ± 0.004 | 0.27 ± 0.003 |
| MT | ✓ | - | - | - | ✓ | - | - | 0.67 ± 0.004 | 0.34 ± 0.001 | 0.62 ± 0.003 | 0.26 ± 0.002 |
| | - | ✓ | - | - | ✓ | - | - | 0.66 ± 0.003 | 0.34 ± 0.004 | 0.63 ± 0.002 | 0.26 ± 0.001 |
| | - | - | ✓ | - | ✓ | - | - | 0.67 ± 0.004 | 0.37 ± 0.002 | 0.64 ± 0.003 | 0.28 ± 0.004 |
| | - | - | - | ✓ | ✓ | - | - | 0.68 ± 0.001 | 0.35 ± 0.003 | 0.65 ± 0.002 | 0.26 ± 0.001 |
| | ✓ | - | - | - | - | ✓ | - | 0.68 ± 0.003 | 0.37 ± 0.004 | 0.65 ± 0.001 | 0.28 ± 0.003 |
| | - | ✓ | - | - | - | ✓ | - | 0.70 ± 0.002 | 0.38 ± 0.001 | 0.67 ± 0.003 | 0.29 ± 0.002 |
| | - | - | ✓ | - | - | ✓ | - | 0.70 ± 0.001 | 0.39 ± 0.003 | 0.67 ± 0.004 | 0.31 ± 0.001 |
| | - | - | - | ✓ | - | ✓ | - | 0.72 ± 0.004 | 0.40 ± 0.002 | 0.70 ± 0.003 | 0.30 ± 0.001 |
| | ✓ | - | - | - | - | ✓ | ✓ | 0.73 ± 0.001 | 0.37 ± 0.002 | 0.70 ± 0.003 | 0.28 ± 0.004 |
| | - | ✓ | - | - | - | ✓ | ✓ | 0.72 ± 0.004 | 0.37 ± 0.001 | 0.70 ± 0.002 | 0.29 ± 0.003 |
| | - | - | ✓ | - | - | ✓ | ✓ | 0.72 ± 0.003 | 0.38 ± 0.004 | 0.69 ± 0.001 | 0.30 ± 0.002 |
| | - | - | - | ✓ | - | ✓ | ✓ | 0.72 ± 0.002 | 0.37 ± 0.003 | 0.68 ± 0.002 | 0.28 ± 0.001 |

We observed a significant performance gain when using BERT features compared to ECFP, as highlighted in Table 2. Figure 1 compares weighted and non-weighted loss functions for both frozen and fine-tuned BERT models. For the frozen BERT model, we observed a consistent performance improvement in balanced accuracy, F1-score, ROC-AUC, and average precision across all modalities when transitioning from Binary Cross-Entropy (BCE) to weighted Binary Cross-Entropy (BCE-w), Focal Loss (FL), and weighted Focal Loss (FL-w). However, this trend was not consistent in the fine-tuned BERT model. Fine-tuning BERT improved performance in some modalities but decreased it in others. In conclusion, BERT features outperformed ECFP, weighted loss functions were superior to unweighted ones, and the effectiveness of fine-tuning remained inconclusive.

To analyze the impact of different loss functions, we computed the log-loss for positive and negative instances generated from Frozen-BERT, as shown in Fig. 2. Task-wise means, calculated using Eq. 10, are represented by small blobs with ellipses indicating the 95% confidence interval across all tasks. The visualization highlighted a network bias towards the majority class (negatives), resulting in elevated log-loss for positive instances, particularly with Binary Cross-Entropy (BCE). Transitioning to weighted Binary Cross-Entropy (BCE-W) prompted the model to equally prioritize both positives and negatives, decreasing log-loss for positives compared to BCE. Moreover, Focal Loss naturally emphasizes on hard examples, in this case, positive instances, lead to significantly lower log-loss for positives. Weighted Focal Loss further supported this by assigning additional weight to positive examples, consequently reducing the log-loss of positive instances even further. Further, in our experience frozen-BERT model provided the highest ROC-AUC with weighted Focal loss as depicted in the bottom row of the Fig. 2.

We computed a task-wise performance, as depicted in Fig. 3. Our models achieved the highest ROC-AUC for biochemistry related tasks and lowest for the clinical tasks the highest ROC scores. Interestingly, the model encounters difficulty in learning two specific tasks: Extramedullary from the pathological category and 100197554 from the clinical category.

## 4   Supplementary Information

The related supplementary information can be found on project GitHub repository https://github.com/Arslan-Masood/Tox_balance

**Acknowledgments.** The authors acknowledge financial support from the European Union's Horizon 2020 research and innovation program under the Marie Skłodowska-Curie grant agreement No 956832, "Advanced Machine learning for Innovative Drug Discovery" (AIDD).

**Disclosure of Interests.** The authors have no competing interests to declare that are relevant to the content of this article.

# References

Ahmad, W., Simon, E., Chithrananda, S., Grand, G. and Ramsundar, B.: ChemBERTa-2: Towards chemical foundation models. arXiv:2209.01712 (2022)

Ai, H., et al.: Predicting drug-induced liver injury using ensemble learning methods and molecular fingerprints. Toxicol. Sci. **165**(1), 100–107 (2018). ISSN 1096-6080, 1096-0929. https://doi.org/10.1093/toxsci/kfy121, https://academic.oup.com/toxsci/article/165/1/100/5000032

Alashwal, H., Lucman, J.: Utilizing cost-sensitive machine learning classifiers to identify compounds that inhibit Alzheimer's APP translation. In: Proceedings of the 2020 4th International Conference on Cloud and Big Data Computing, pp. 113–117, Virtual United Kingdom. ACM (2020). ISBN 978-1-4503-7538-2. https://doi.org/10.1145/3416921.3416931, https://dl.acm.org/doi/10.1145/3416921.3416931

Asilar, E., Hemmerich, J., Ecker, G.F.: Image based liver toxicity prediction. J. Chem. Inform. Model. **60**(3), 1111–1121 (2020). ISSN 1549-9596, 1549-960X. https://doi.org/10.1021/acs.jcim.9b00713, https://pubs.acs.org/doi/10.1021/acs.jcim.9b00713

Chawla, N.V., Bowyer, K.W., Hall, L.O., Kegelmeyer, W.P.: SMOTE: Synthetic Minority Over-sampling Technique. J. Artif. Intell. Res. **16**, 321–357 (2002). ISSN 1076-9757. https://doi.org/10.1613/jair.953, https://www.jair.org/index.php/jair/article/view/10302

Chen, M., Borlak, J., Tong, W.: High lipophilicity and high daily dose of oral medications are associated with significant risk for drug-induced liver injury. Hepatology, **58**(1), 388–396 (2013). ISSN 02709139. https://doi.org/10.1002/hep.26208, https://onlinelibrary.wiley.com/doi/10.1002/hep.26208

Chen, M., et al. Quantitative structure-activity relationship models for predicting drug-induced liver injury based on FDA-approved drug labeling annotation and using a large collection of drugs. Toxicol. Sci. **136**(1), 242–249 (2013). ISSN 1096-6080, 1096-0929. https://doi.org/10.1093/toxsci/kft189, https://academic.oup.com/toxsci/article-lookup/doi/10.1093/toxsci/kft189

Chithrananda, S., Grand, G. and Ramsundar, B.: ChemBERTa: Large-scale self-supervised pretraining for molecular property prediction. arXiv preprint arXiv:2010.09885 (2020)

Cruz-Monteagudo, M., Cordeiro, M.N.D., Borges, F.: Computational chemistry approach for the early detection of drug-induced idiosyncratic liver toxicity: early Detection of Drug-Induced Idiosyncratic Liver Toxicity. Jo. Comput. Chem. **29**(4), 533–549 (2008.) ISSN 01928651. https://doi.org/10.1002/jcc.20812, https://onlinelibrary.wiley.com/doi/10.1002/jcc.20812

Davis, J., Goadrich, M.: The relationship between precision-recall and ROC curves. In: Proceedings of the 23rd international conference on Machine learning, ICML 2006, pp. 233–240, New York, NY, USA. Association for Computing Machinery (2006). ISBN 978-1-59593-383-6. https://doi.org/10.1145/1143844.1143874, https://doi.org/10.1145/1143844.1143874

Devlin, J., Chang, M.-W., Lee, K., Toutanova, K.: BERT: pre-training of deep bidirectional transformers for language understanding. arXiv:1810.04805 (2019)

Eitrich, T., Kless, A., Druska, C., Meyer, W., Grotendorst, J.: Classification of highly unbalanced CYP450 data of drugs using cost sensitive machine learning techniques. J. Chem. Inform. Model. **47**(1), 92–103 (2007). ISSN 1549-9596. https://doi.org/10.1021/ci6002619, https://doi.org/10.1021/ci6002619. Publisher: American Chemical Society

Ekins, S., Williams, A.J., Xu, J.J.: A predictive ligand-based bayesian model for human drug-induced liver injury. Drug Metab. Dispos. **38**(12), 2302–2308 (2010). ISSN 0090-9556, 1521-009X. https://doi.org/10.1124/dmd.110.035113, http://dmd.aspetjournals.org/lookup/doi/10.1124/dmd.110.035113

Elkan, C.: The foundations of cost-sensitive learning. In: Proceedings of the 17th International Joint Conference on Artificial Intelligence - Volume 2, IJCAI 2001, pp. 973–978, San Francisco, CA, USA. Morgan Kaufmann Publishers Inc.(2001). ISBN 978-1-55860-812-2

Fabian, B., et al.: Molecular representation learning with language models and domain-relevant auxiliary tasks. arXiv:2011.13230 (2020)

Forman, G., Scholz, M.: Apples-to-apples in cross-validation studies: pitfalls in classifier performance measurement. ACM SIGKDD Explor. Newslett. **12**(1), 49–57 (2010). ISSN 1931-0145, 1931-0153. https://doi.org/10.1145/1882471.1882479, https://dl.acm.org/doi/10.1145/1882471.1882479

Gao, K., Nguyen, D.D., Sresht, V., Mathiowetz, A.M., Tu, M., Wei, G.W.: Are 2D fingerprints still valuable for drug discovery? Phys. Chem. Chem. Phys. **22**(16), 8373–8390 (2020). ISSN 1463-9076, 1463-9084. https://doi.org/10.1039/D0CP00305K, http://xlink.rsc.org/?DOI=D0CP00305K

Goodfellow, I., et al.: Generative adversarial nets. In: Advances in Neural Information Processing Systems, vol. 27 (2014)

Harnik, Y., Milo, A.: A focus on molecular representation learning for the prediction of chemical properties. Chem. Sci. **15**(14), 5052–5055 (2024). ISSN 2041-6520, 2041-6539. https://doi.org/10.1039/D4SC90043J, https://xlink.rsc.org/?DOI=D4SC90043J

Igarashi, Y., et al.: Open TG-GATEs: a large-scale toxicogenomics database. Nucleic Acids Res. **43**(D1), D921–D927 (2015). ISSN 1362-4962, 0305-1048. https://doi.org/10.1093/nar/gku955, https://academic.oup.com/nar/article/43/D1/D921/2439524

Irwin, R., Dimitriadis, S., He, J., Bjerrum, E.J.: Chemformer: a pre-trained transformer for computational chemistry. Mach. Learn. Sci. Technol. **3**(1), 015022 (2022). ISSN 2632-2153. https://doi.org/10.1088/2632-2153/ac3ffb, https://dx.doi.org/10.1088/2632-2153/ac3ffb. Publisher: IOP Publishing

Jain, S., et al.: Large-scale modeling of multispecies acute toxicity end points using consensus of multitask deep learning methods. J. Chem. Inform. Model. **61**(2), 653–663 (2021). ISSN 1549-9596, 1549-960X. https://doi.org/10.1021/acs.jcim.0c01164, https://pubs.acs.org/doi/10.1021/acs.jcim.0c01164

Kim, J.H., Scialli, A.R.: Thalidomide: the tragedy of birth defects and the effective treatment of disease. Toxicol. Sci. **122**(1), 1–6 (2011). ISSN 1096-6080, 1096-0929. https://doi.org/10.1093/toxsci/kfr088, https://academic.oup.com/toxsci/article/1672454/Thalidomide:

Kingma D.P., Welling, M.: Auto-encoding variational bayes. arXiv preprint arXiv:1312.6114 (2013)

Kuhn, M., Letunic, I., Jensen, L.J., Bork, P.: The SIDER database of drugs and side effects. Nucleic Acids Res. **44**(D1), D1075–D1079 (2016) ISSN 0305-1048, 1362-4962. https://doi.org/10.1093/nar/gkv1075, https://academic.oup.com/nar/article-lookup/doi/10.1093/nar/gkv1075

Lancaster, E.M., Hiatt, J.R., Zarrinpar, A.: Acetaminophen hepatotoxicity: an updated review. Arch. Toxicol. **89**, 193–199 (2014). https://doi.org/10.1007/s00204-014-1432-2

Laveti, R.N., Mane, A.A., Pal, S.N.: Dynamic stacked ensemble with entropy based undersampling for the detection of fraudulent transactions. In: 2021 6th International Conference for Convergence in Technology (I2CT), pp. 1–7, Maharashtra,

India. IEEE (2021). ISBN 978-1-72818-876-8. https://doi.org/10.1109/I2CT51068. 2021.9417896, https://ieeexplore.ieee.org/document/9417896/

Lee, W., Seo, K.: Downsampling for binary classification with a highly imbalanced dataset using active learning. Big Data Res. **28**, 100314 (2022). ISSN 22145796. https://doi.org/10.1016/j.bdr.2022.100314, https://linkinghub.elsevier.com/retrieve/pii/S2214579622000089

Li, J., Jiang, X.: Mol-BERT: an effective molecular representation with BERT for molecular property prediction. Wireless Commun. Mob. Comput. **2021**, 1–7 (2021). ISSN 1530-8677, 1530-8669. https://doi.org/10.1155/2021/7181815, https://www.hindawi.com/journals/wcmc/2021/7181815/

Li, X., Fourches, D.: Inductive transfer learning for molecular activity prediction: *Next-Gen QSAR Models with MolPMoFiT*. J. Cheminform. **12**(1), 1–15 (2020). https://doi.org/10.1186/s13321-020-00430-x

Li, Z., Jiang, M., Wang, S., Zhang, S.: EEP learning methods for molecular representation and property prediction. Drug Discov. Today **27**(12), 103373 (2022). ISSN 1878-5832. https://doi.org/10.1016/j.drudis.2022.103373

Lin, T.Y., Goyal, P., Girshick, R., He, K., Dollár, P.: Focal loss for dense object detection. arXiv:1708.02002 (2018)

Liu, Y., Zhang, R., Li, T., Jiang, J., Ma, J., Wang, P.: MolRoPE-BERT: an enhanced molecular representation with rotary position embedding for molecular property prediction. J. Mol. Graph. Model. **118**, 8344 (2023) ISSN 1093-3263. https://doi.org/10.1016/j.jmgm.2022.108344, https://www.sciencedirect.com/science/article/pii/S1093326322002236

López, V., Fernández, A., Moreno-Torres, J.G., Herrera, F.: Analysis of preprocessing vs. cost-sensitive learning for imbalanced classification. Open problems on intrinsic data characteristics. Expert Systems with Applications, **39**(7), 6585–6608 (2012). ISSN 09574174. https://doi.org/10.1016/j.eswa.2011.12.043, https://linkinghub.elsevier.com/retrieve/pii/S0957417411017143

Maziarka, Ł., Danel, T., Mucha, S., Rataj, K., Tabor, J., Jastrzębski, S.: Molecule attention transformer. arXiv:2002.08264 (2020)

Moein, M., et al.: Chemistry-based modeling on phenotype-based drug-induced liver injury annotation: from public to proprietary data. Chem. Res. Toxicol. **36**(8), 1238–1247 (2023). ISSN 0893-228X, 1520-5010. https://doi.org/10.1021/acs.chemrestox.2c00378, https://pubs.acs.org/doi/10.1021/acs.chemrestox.2c00378

Parasrampuria, D.A., Benet, L.Z., Sharma, A.: Why drugs fail in late stages of development: case study analyses from the last decade and recommendations. AAPS J **20**(3), 1–16 (2018). https://doi.org/10.1208/s12248-018-0204-y

Singh Rawat, S., Mishra, A.K.: Review of methods for handling class-imbalanced in classification problems. arXiv:2211.05456 (2022)

Rezende, D., Mohamed, S.: Variational inference with normalizing flows. In: International Conference on Machine Learning, pp. 1530–1538. PMLR (2015)

Rogers, D., Hahn, M.: Extended-connectivity fingerprints. J. Chem. Inform. Model. **50**(5), 742–754 ) (2010). ISSN 1549-9596, 1549-960X. https://doi.org/10.1021/ci100050t, https://pubs.acs.org/doi/10.1021/ci100050t

Ruthotto, L., Haber, E.: An introduction to deep generative modeling (2021)

Sharma, B., et al.: Accurate clinical toxicity prediction using multi-task deep neural nets and contrastive molecular explanations. Sci. Rep. **13**(1), 4908 (2023). ISSN 2045-2322. https://doi.org/10.1038/s41598-023-31169-8, https://www.nature.com/articles/s41598-023-31169-8

Shermukhamedov, S., Mamurjonova, D., Probst, M.: Structure to property: chemical element embeddings and a deep learning approach for accurate prediction of chemical properties arXiv:2309.09355 (2023)

Sosnin, S., Karlov, D., Tetko, I.V., Fedorov, M.V.: Comparative study of multitask toxicity modeling on a broad chemical space. J. Chem. Inform. Model. **59**(3), 1062–1072 (2019). ISSN 1549-9596, 1549-960X. https://doi.org/10.1021/acs.jcim.8b00685, https://pubs.acs.org/doi/10.1021/acs.jcim.8b00685

Stanley, L.A.: Chapter 27 - Drug Metabolism. In: Badal, S., Delgoda, R., (eds.) Pharmacognosy, pp. 527–545. Academic Press, Boston (2017). ISBN 978-0-12-802104-0. https://doi.org/10.1016/B978-0-12-802104-0.00027-5, https://www.sciencedirect.com/science/article/pii/B9780128021040000275

Wang, Y., Xiao, Q., Chen, P., Wang, B: In silico prediction of drug-induced liver injury based on ensemble classifier method. Int. J. Mol. Sci. **20**(17), 4106 (2019). ISSN 1422-0067. https://doi.org/10.3390/ijms20174106, https://www.mdpi.com/1422-0067/20/17/4106

Wang, H., Liu, R., Schyman, P., Wallqvist, A.: Deep neural network models for predicting chemically induced liver toxicity endpoints from transcriptomic responses. Front. Pharmacol. **10**, 42 (2019). ISSN 1663-9812. https://doi.org/10.3389/fphar.2019.00042, https://www.frontiersin.org/article/10.3389/fphar.2019.00042/full

Wang, S., Guo, Y., Wang, Y., Sun, H., Huang, J.: SMILES-BERT: large scale unsupervised pre-training for molecular property prediction. In: Proceedings of the 10th ACM International Conference on Bioinformatics, Computational Biology and Health Informatics, pp. 429–436, Niagara Falls NY USA (2019). ACM. ISBN 978-1-4503-6666-3. https://doi.org/10.1145/3307339.3342186, https://dl.acm.org/doi/10.1145/3307339.3342186

Watkins, P.B.: Drug safety sciences and the bottleneck in drug development. Clin. Pharmacol. Ther. **89**(6), 788–790 (2011). ISSN 0009-9236, 1532-6535. https://doi.org/10.1038/clpt.2011.63, https://onlinelibrary.wiley.com/doi/10.1038/clpt.2011.63

Weininger, D.: SMILES, a chemical language and information system. 1. Introduction to methodology and encoding rules. J. Chem. Inform. Comput. Sci. **28**(1), 31–36 (1988). ISSN 0095-2338. https://doi.org/10.1021/ci00057a005, https://doi.org/10.1021/ci00057a005. Publisher: American Chemical Society

Williams, D.P., Lazic, S.E., Foster, A.J., Semenova, E., Morgan, P.: Predicting drug-induced liver injury with Bayesian machine learning. Chem. Res. Toxicol **33**(1), 239–248 (2020). ISSN 0893-228X, 1520-5010. https://doi.org/10.1021/acs.chemrestox.9b00264, https://pubs.acs.org/doi/10.1021/acs.chemrestox.9b00264

Winter, R., Montanari, F., Noé, F., Clevert, D.A.: Learning continuous and data-driven molecular descriptors by translating equivalent chemical representations. Chem. Sci. **10**(6), 1692–1701 (2019). ISSN 2041-6520, 2041-6539. https://doi.org/10.1039/C8SC04175J, https://xlink.rsc.org/?DOI=C8SC04175J

Xu, Y., Dai, Z., Chen, F., Gao, S., Pei, J., Lai, L., Deep learning for drug-induced liver injury. J. Chem. Inform. Model. **55**(10), 2085–2093 (2015). ISSN 1549-9596, 1549-960X. https://doi.org/10.1021/acs.jcim.5b00238, https://pubs.acs.org/doi/10.1021/acs.jcim.5b00238

Yang, K., et al.: Analyzing learned molecular representations for property prediction. arXiv:1904.01561 (2019)

Zhu, M.: Recall, precision and average precision. Department of Statistics and Actuarial Science, University of Waterloo, Waterloo, vol. 2, no. 30, p. 6 (2004)

**Open Access** This chapter is licensed under the terms of the Creative Commons Attribution 4.0 International License (http://creativecommons.org/licenses/by/4.0/), which permits use, sharing, adaptation, distribution and reproduction in any medium or format, as long as you give appropriate credit to the original author(s) and the source, provide a link to the Creative Commons license and indicate if changes were made.

The images or other third party material in this chapter are included in the chapter's Creative Commons license, unless indicated otherwise in a credit line to the material. If material is not included in the chapter's Creative Commons license and your intended use is not permitted by statutory regulation or exceeds the permitted use, you will need to obtain permission directly from the copyright holder.

# Registries in Machine Learning-Based Drug Discovery: A Shortcut to Code Reuse

Peter B. R. Hartog[1,3]($\boxtimes$) , Emma Svensson[2,3]($\boxtimes$) , Lewis Mervin[4] ,
Samuel Genheden[3] , Ola Engkvist[3,5] , and Igor V. Tetko[1]

[1] Institute of Structural Biology, Molecular Targets and Therapeutics Center, Helmholtz Munich-Deutsches Forschungszentrum für Gesundheit und Umwelt (GmbH), 58764 Neuherberg, Germany
[2] ELLIS Unit Linz, Institute for Machine Learning, Johannes Kepler University Linz, 4040 Linz, Austria
[3] Molecular AI, Discovery Sciences, R&D, AstraZeneca Gothenburg, 431 83 Gothenburg, Sweden
{peter.hartog,emma.svensson1}@astrazeneca.com
[4] Molecular AI, Discovery Sciences, R&D, AstraZeneca Cambridge, Cambridge CB2 0AA, UK
[5] Department of Computer Science and Engineering, Chalmers University of Technology, 412 96 Gothenburg, Sweden

**Abstract.** Computer-aided drug discovery gradually builds on previous work and requires reusable code to advance research. Currently, research code is mainly used to provide further insights into the original research whilst code reuse has a lower priority. Modularity, the segmentation of code for independent modules, promotes good coding practices and code reuse. The registry pattern has been proposed as a way to call functionalities dynamically, but it is currently overlooked as a shortcut to promote code reuse. In this work, we expand the registry pattern to better suit computer-aided drug discovery and achieve a unified, reusable, and interchangeable interface with optional meta information. Our reformulated pattern is particularly suitable for collaborative research with standardized frameworks where multiple internal and external modules are used interchangeably and coding is more focused on fast iteration over low-debt technical code, such as in machine learning-based research for drug discovery. In a workflow, we exemplify the usage of the design patterns. Additionally, we provide two case studies where we 1) showcase the effectiveness of registration in a larger collaborative research group, and 2) overview the potential of registration in currently available open-source tools. Finally, we empirically evaluate the registry pattern through previous implementations and indicate where additional functionality can improve its use.

**Keywords:** registration · design pattern · modularity · code reuse · drug discovery · machine learning

© The Author(s) 2025
D.-A. Clevert et al. (Eds.): AIDD 2024, LNCS 14894, pp. 98–115, 2025.
https://doi.org/10.1007/978-3-031-72381-0_9

# 1   Introduction

The development of computer-aided drug discovery relies on previous research from multiple fields to bridge the knowledge gap between domain experts and computer scientists [41]. As such, software development in the field is often built up of a combination of open-source tools, collaborative developments, and independent research. Currently, research code is mainly used to provide further insights into the original research rather than to use in future research [6]. There is also a reproducibility problem [3] of code that stems from the low priority of code reuse, as noted by Nature Computational Science ("But is the code (re)usable?" [Editorial]. 2021, 23 July). Benureau and Rougier [5] proposed that research code should adhere to particular requirements for stable and reliable results. Code should be replicable, to obtain the same results as the original paper; be able to run without problems; repeatable (i.e., deterministic); reproducible (i.e., deterministic over multiple runs); and, finally, reusable. However, research code is different from production software because its goals are focused mostly on replication, where reuse is often an afterthought.

Code reuse is dependent on the concept of modularity [4]. Modular code is code that is grouped with related code and mostly independent of other parts of the code, named high cohesion and low coupling, respectively [33,46]. This results in code that is interchangeable, replaceable, and can be updated and used without issue [23,25,39]. Modular code also avoids the need to repeat code segments [17] and forces code modules to achieve single objectives over multiple responsibilities [26].

Machine learning (ML) workflows are inherently modular (Fig. 1). The workflow of ML is usually segmented into separate steps, such as data generation and model creation, regardless of implementation. ML approaches are becoming more prominent for research in drug discovery [7,11,42]. Computational research, like all scientific research, builds on the knowledge from previous discoveries and utilizes known methods and coding frameworks to create new tools, apply methods to new fields, or investigate new problems. ML researchers use the available tools to compare their novel models to previous approaches [9,22,45] and to streamline their pipeline [14,47]. There are also tools designed explicitly for ML in drug discovery, some of which focus on aiding new practitioners in quickly finding and comparing state-of-the-art approaches [10,24,28,34,40,44]. Other similar tools can function to bridge the knowledge gap between natural scientists and computer scientists. The latter can be achieved either by supplying domain-specific knowledge to ML workflows [16,21,27,31] or by making common ML frameworks [1,32] more accessible through higher-level abstractions [8,15,43]. Although tools are created to be used by others, and therefore surpass research code in reusability, tools are often semi-rigid, made to work out of the box for a fixed domain, purpose, or type of method. This means there is often a high bar to add new functionalities in open-source tools [4].

In this work, we identify registries [12] as a shortcut to code reuse for ML-based drug discovery. We introduce the registry design pattern to those unfamiliar and suggest additional capabilities. Furthermore, we identify situations where

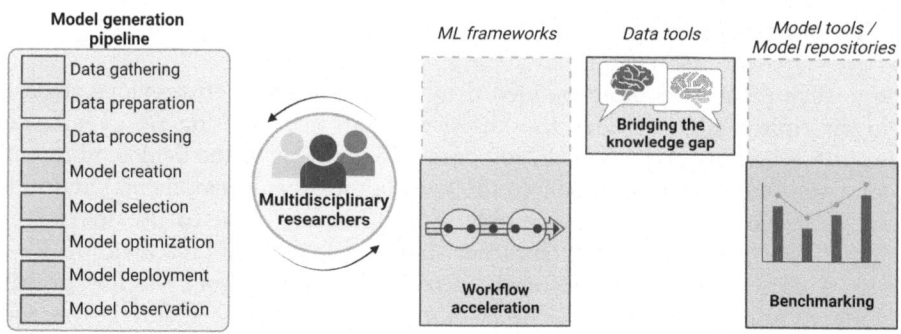

**Fig. 1. Overview of model generation pipeline and applicability of tools for each step. Repository tools** are primarily used for reproducibility and to benchmark new approaches against the existing state-of-the-art. **Data tools** bridge the knowledge gap between domain experts and ML practitioners. Finally, **ML frameworks** add workflow abstractions that accelerate the ML pipeline. Multidisciplinary collaborations rely on the use of a combination of tools from these categories, as illustrated in the leftmost panel.

the cheminformatics community can benefit from registries to easily make their code more reusable as well as more replicable [5]. We provide an open-source implementation of the generic registry design pattern through the Python Package Index. While the fast pace of drug discovery research can result in bad coding practices, the simplicity of our proposed registry is meant to encourage improved coding practices with minimal effort for the researchers. The use of registries during codebase design can additionally serve as an easy way to enforce desired behaviors, such as a factory pattern or test adherence, which in turn helps contributors adhere to the desired coding standards. Finally, we identify where previous implementations of registries have succeeded and failed, and discuss the reasons behind successful implementations in both research code and software tools. Our contributions are summarized as follows.

- We extend previous explanations of the registry with capabilities for use by researchers.
- We outline several use cases of where and when registries can be a shortcut for reuse in computer-aided drug discovery research.
- We identify previous implementations of registries and note their positive and negative implications.

## 2    Methods

The registry design pattern has been proposed as a tool for dynamic instance creation of object-oriented classes [12]. Best practices in object-oriented programming are often formalized as design patterns. A software design pattern provides a template for a general and reusable way that solves a recurring problem in software engineering [13]. We reformulate the registry as a method to retrieve similar modules with similar functionalities or uses. Consider a set of modules $\mathcal{M}$ supplying the same type of functionality but with different implications to a workflow. Given that the modules follow the Liskov substitution principle [23], i.e. that $f : \mathcal{X} \rightarrow \mathcal{Y}, \forall f \in \mathcal{M}$ where $\mathcal{X}$ and $\mathcal{Y}$ are the set of inputs and outputs respectively, it is commonly known that they can be used interchangeably through inheritance. A problem with inheritance is that each module still has to be initialized individually. The standard design pattern to interchange such modules is the factory pattern (Fig. 2 left). However, this approach requires strict inheritance from an abstract class.

The registry design pattern uses call and set functions, usually renamed to get and register, to dynamically set and retrieve objects from a unified storage location. Registry systems are often initialized at run-time and used in combination with alias-based retrieval. As such, a registry follows the factory design pattern in that it provides a common interface for categorically similar functionalities without the explicit need for concrete classes. Furthermore, the registry encapsulates each alternative module, hiding individual details behind a unified set of function calls.

Formally, we reformulate the existing registry design pattern [12] as a means to collect interchangeable modules with encapsulated functionalities retrievable using a unified command. Additionally, the registry pattern is dynamic in its application. We provide our proposed design pattern to the cheminformatics community through the Python Package Index as the registry-factory package under the MIT license [29], which specifies the implementation of a collection of registries. The open-source code is available at https://github.com/aidd-msca/registry-factory.

## 3   Results

### 3.1   Workflow: Creating a Registry and Registering Modules

An overview of the process of creating a registry is illustrated in Fig. 3. A new registry is either imported from the package or is instantiated from a factory class. No instance of a registry needs to be created to use it. First, a section of code is separated from the framework. This is then converted into a function or class and registered into the registry. All subsequent scripts and external collaborators are then able to call upon the registry for this code.

When a registry is outlined as above it promotes and allows specific actions to be performed more fluently: 1) A registry provides a framework to exchange modules with similar functions. This switch is also stable in execution and flexible in application, depending on how strictly the registry is set up. 2) The registry setup allows control over how modules should be set up. More strict setups will force standardized modularity and interfaces, whilst more flexible setups allow faster extension and broader application. 3) Due to the standardization, more internal and external modules can use the same execution framework to function.

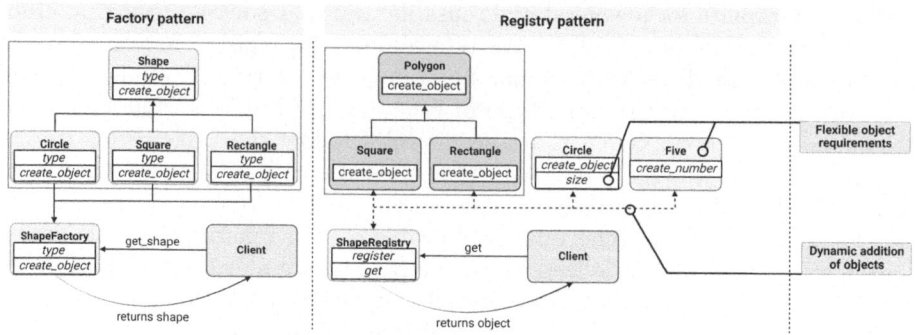

**Fig. 2. Unified Modeling Language (UML) diagram of the factory design pattern and our proposed Registry design pattern. Left:** UML of the factory pattern. The ShapeFactory functions as an interface where all subclasses of the Shape superclass can be called by the client. **Right:** UML of the registry pattern. The ShapeRegistry functions as an interface where any object no matter their superclass can be registered dynamically and called by the client.

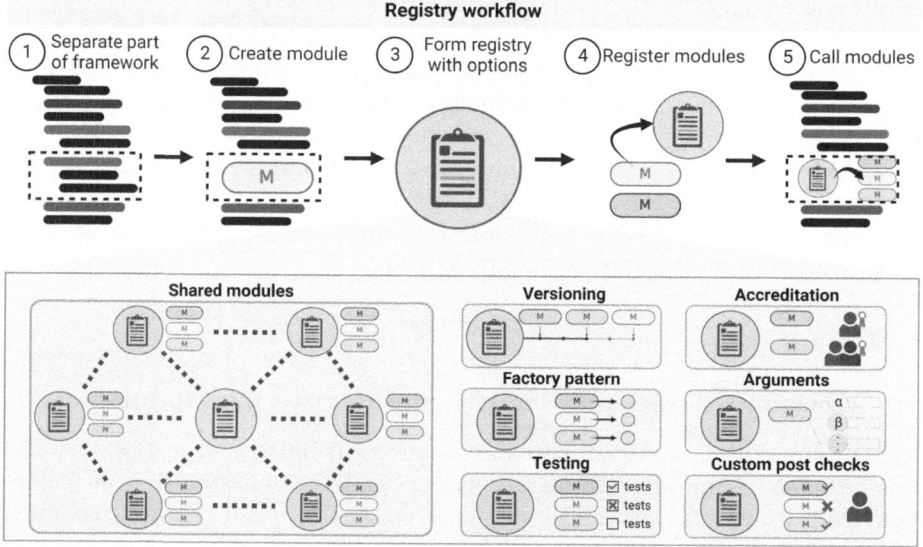

**Fig. 3. Workflow of creating and using a registry.** 1) Part of the code framework is identified which can be separated from the rest. 2) This section is modularized to be independent of the rest of the code. 3) A registry is created. Here, the choice of additional meta information such as versioning, accreditation, and arguments are set as well as the choice to share modules and force a specific class pattern. There is also the option to add post-registration checks, both custom and those that reference a testing script. 4) New modules are registered to the registry. 5) Finally, the modules are called in the main workflow using the registry.

## 3.2    Enhanced Functionality and Expanded Capabilities

Minimal usage of registries can be limited in their application. As a result, additional capabilities can increase the application options of registries, even in more advanced architecture designs. In this paper, we advocate for a non-exhaustive selection of additional capabilities and have implemented them into a separate package that will be made available upon acceptance (Fig. 3).

Upon creation, the specifications of the registry are instantiated and a selection of additional capabilities can be included. Firstly, to increase ease of use, inter-registry module sharing and argument registration are implemented. Secondly, the following features are implemented to allow for better control in a codebase setting, factory pattern forcing, versioning, and automatic testing upon registration. Finally, to accommodate the research community when contributing to packages, we have implemented accreditation which can be retrieved when calling registered objects.

## 3.3    A Shortcut to Modularity and Reusability

The main reason for the integration of registries is the passive enforcement of modularity. Experienced programmers will inherently shift to modular code and separate classes and functions into coherent modules and packages. However, research code and practitioners from other fields in the cheminformatics community will often focus on fast iteration over low-debt technical code. As such, the use of registries aims to passively enforce the usability of sections that the researcher will need repeatedly without the need for that code to be of high standard. Similarly, registries allow researchers to share code more easily, both internally during a project and externally after the research has been published. Contrary to common Python principles the use of registries gives a more flexible way of sharing code such that certain parts can be reused while others are updated or changed entirely. For example, this can be achieved by using the registries as hooks, allowing researchers to add functionalities without altering the code.

**Case Study: Codebase Design and Collaboration.** Registries are ideal for collaborative work, as they signify which part of the code is reusable to collaborators. The field of cheminformatics is multidisciplinary by definition and researchers often work collaboratively or using shared codebases. In these settings, low-quality code can prohibit collaboration.

By utilizing registries in collaborative work, researchers can specify beforehand what parts of code can be easily shared between collaborators. It also stops collaborators from having to dive into messy code and instead be able to just extract the segments of interest.

Furthermore, codebase designers and maintainers can use registries together with our added capabilities to automatically check and control suggested code submissions. Using automatic testing, registered modules are submitted to testing upon entering the registry, whereas custom controls can enforce desired behaviors, such as consistent input variables. This approach not only eliminates redundancy but also enhances code readability and maintainability.

**Case Study: Registries in Cheminformatics Tools.** The usage of cheminformatics tools can also benefit from registries. Table 1 gives a non-exhaustive list of available tools often used in ML for drug discovery. These tools are essential in their respective domains but they can be difficult to combine or use interchangeably. Many of them are internally built in a modular way but less so developed to be adjusted by the users. The registry design pattern can be used on different levels together with these tools to create adaptability and interchangeability, which in turn allows for code reuse.

Firstly, including registration can allow the individual tools to open up the possibility for users to contribute with their own functionalities or include other open-source packages. Users can test out new functionalities directly in the tool environment using registries, without the need to download and add to the package code. Model repository tools can benefit from registries by allowing users to register additional models in the collection. One can imagine entire libraries of models and data collections being allowed in ML pipeline tools or workflow systems, such that different collections can be used interchangeably.

Additionally, tools can open up internal capabilities using registries. Oftentimes, modules from cheminformatics tools are built with a specific functionality in mind. However, most modules contain multiple useful functionalities inside which can be used outside of that specific module. The use of registries allows users to easily extract internal capabilities and use them in their own code.

Finally, allowing users access to internal sections allows them to switch parts of the internal characteristics of tools. In the field of ML, this can include functionalities such as custom loss functions, weight initialization schemes, layers, or activation functions. In a more general sense, framework tools can create registries by specifying the sections that can be altered, and controlling how these sections operate in a unified interface.

Consequently, implementation of registries in open-source code allows for quick benchmarks, inherently supports contribution to tools, and promotes code reuse.

## 3.4    Empirical Evaluation: Application and Impact in Previous Implementations

Previous packages have been implemented with versions of the registry that we propose. Here, we assess their impacts and analyze possible limitations in the implementations. Here, we analyze two model repositories and one ML pipeline package that have internal registries, the graph-based TorchDrug [49], the model training GT4SD package [24] and the ML pipeline package MLFlow [47]. We also compare these packages with how the highly cited and often-used Hugging Face package [45] operates. The Hugging Face package uses an online repository system to collect machine learning models and benchmarking datasets. For this, the

**Table 1. Overview of tools.** Non-exhaustive overview of open-source tools used for ML and/or drug discovery.

| Software | Description |
|---|---|
| **Data tools** | |
| CDK [38] | Chemistry development kit with methods for molecular informatics. |
| RDKit [21] | Extensive toolkit for cheminformatics logic and functionalities. |
| OpenBabel [31] | Toolbox with functionalities for chemical languages. |
| TDC [16] | Collection of benchmarks in several drug discovery applications. |
| DataMol [27] | Library for intuitive manipulation of molecules. |
| Datasets [22] | HuggingFace collection of natural language dataset. |
| ... | |
| **Model repositories** | |
| OCHEM [40] | ML framework for the collection of QSAR models. |
| Transformers [45] | HuggingFace collection of language models. |
| bio_embedding [10] | State-of-the-art language models for protein encoding. |
| solo-learn [9] | Collection of self-supervised models for representation learning. |
| GT4SD [24] | Generative modeling environment for material discovery. |
| ... | |
| **ML pipelines** | |
| ODDT [44] | Traditional ML methods applied to drug discovery. |
| TensorFlow [1] | General tool for deep learning logic. |
| MLFlow [47] | Standardized ML workflow. |
| PyTorch [32] | General tool for deep learning logic. |
| DeepChem [34] | High-level ML framework for drug discovery. |
| AMPL [28] | High-level ML framework for drug discovery. |
| MetaFlow [14] | Standardized ML workflow. |
| TorchDrug [49] | Geometric deep learning for drug discovery. |
| ... | |
| **Workflow systems** | |
| KNIME [36] | Graphical user interface for data analytics with components for ML. |
| AZOrange [37] | Graphical environment for high-performance ML-based QSAR models. |
| ... | |

package uses an alias resolver similar to registry calls to map string names to model instances and classes. As such, it has no obvious relation to the registry design pattern but it does use many of the same functionalities. TorchDrug is a package that uses a registry as an alias resolver analog as well, but that also actively uses it for registration purposes. Here, models and datasets are registered to be easily retrieved by users using simple string representations. It then further supports changing models and datasets within their internal pipeline, opening up the interface to other users. In GT4SD, the available algorithms are stored in a registry that can be called upon to retrieve each algorithm. Here, the registry is used to combine the interfaces of different molecular representation and prediction models and collections, including those from TorchDrug and Hugging Face. This is a good example of how to use registries to combine models from different classes. Finally, the MLFlow package uses online registration of models with version tracking and aliasing. It uses registries, but its registry is online or saved to a local log file. This is primarily used for model training, versioning, and benchmarking.

Both TorchDrug and GT4SD use registries internally built in a modular way but are less useful for user adjustments. As mentioned, registries can be used on different levels to create adaptability and interchangeability. TorchDrug uses its registry to allow users to add datasets to their selection and then use these new datasets similarly to their ways, seamlessly integrating new data into the workflow. GT4SD uses its registries more to standardize the interface between the model libraries of other packages. While Hugging Face uses the basic alias call function, but not the registry function itself, it is clear that it values the capabilities of registries, though prefers the higher-level modularity that allows users to publish code in a GitHub fashion over code snippets.

## 4   Discussion

In this work, we have introduced the idea of using the registry design pattern to promote code reuse, as well as other good coding practices. In the following section, we discuss previous adoptions, specify important aspects to consider when employing registries, and outline the advantages and disadvantages of registries implemented for research.

### 4.1   Adoption in Previous Implementations

The three current implementations of registries in packages that we have analyzed, in TorchDrug [49], GT4SD [24] and MLFlow [47], indicate promise usage of registries, but these implementations lack the simple integration of opening up internal modules to changes and only use it to change external models. Hugging Face mentions as much in their description of the Transformers package [45], where models are exclusively used for comparison and simple optimization, not for further refinement. This means that the registries, or registry-like systems, are limited in their applicability. GT4SD does something more interesting, in that it uses its registry to combine the different registered objects from both Hugging Face and TorchDrug. They achieve a large library of models however, it means that the package is somewhat limited in its integration of further models by users. Both TorchDrug and GT4SD have the issue that registries are mostly used internally to resolve and gather different models, rather than a method of code reuse. This can limit external contributions to the package. We also note that models are more often submitted to Hugging Face compared to packages whose registry systems are more code-based rather than online. One of the reasons might be that Hugging Face has a very clear way and tutorials regarding how to contribute to the Transformers package, as well as allowing local integration. A second reason for the mismatch between TorchDrug, GT4SD, and Hugging Face is the broader view as well as the adoption of Hugging Face as a platform, meaning that a critical mass may have been achieved for the Hugging Face package that promotes registering models there over other systems. However, note that TorchDrug also has a significant amount of external contributions and users of more than the implemented models.

## 4.2   Registration as a Design Pattern

Design patterns are software generic solutions to problems that often arise [13]. When registration is implemented at the start of the project, it enforces modular code. If it is instead adopted into preexisting code, it allows users to use any standardized framework and switch a segment out to replace it with external code. There is an ongoing debate on the effectiveness of design patterns in general, criticizing the relative lack of empirical evidence of effectiveness [2,48]. However, meta-studies conclude that the original design patterns [13] are mostly correlated with system complexity [19], which in turn is positively correlated with system design quality [18]. This leads to the suggestion that registries as a design pattern might be best used in complex systems, or, in the case of ML for drug discovery, in highly consistent systems where the calls to the registry are sparse.

Jaspan et al. have found that coding speed depends on code visibility [20]. Due to the ability to register any object using registries, registration introduces some encapsulation and information hiding because it makes major module logic inaccessible from the execution program. Despite promoting good software design principles, including decoupling [30], this also introduces a layer of invisibility for the programmer. Moreover, the layer of invisibility can impede speed by making it harder to track source code. However, the main advantage of encapsulation is the ease of interchanging modules without knowledge about the encapsulated function. Therefore, the trade-off is the need to inspect the internal logic of encapsulated modules, and researchers should consider this when deciding which objects to keep in a registry.

### 4.3    Advantages and Considerations of the Registry

Our reformulation of the registry design pattern offers several advantages for software development, including increased modularity and reusability, improved interchangeability, enhanced code clarity, increased stability, and ease of future extensions. In the following section, these advantages and considerations are discussed in more detail.

**Modularity.** Registration allows researchers to easily list and call objects to standardize workflows and switch out modules in a structured and flexible manner. Researchers can register any object, from small modular functionalities to entire scripts. As a result, total flexibility in the scale of modularity is possible, which crucially also allows the integration of external tools. The dependency of registration on modularity is through their execution mechanism where modules require similar input-output regimens, which inherently pushes for modular design choices in research code. There are two things to consider when making modular design choices.

The first consideration for modular design is composability. The composability of a system refers to the relationships between modules in the execution. Following the classification described by Sarjoughian et al. [35], modules can be composed to follow one another hierarchically (*mono*) or be used within high-level modules (*super*). Additionally, there are the *meta* and *poly* composability options. These options are higher-level systems to translate the *mono* system, where the execution runs on the transformed modules in the higher-level system. Due to their complexity, these latter composability options are often avoided in ML. The use of registries helps to make the composability clear to anyone using the system and spells out what can and should be modified.

The second consideration is that modularity can be coded on different scales of abstraction. Module abstraction describes the scope that a module has. Code can be modular on a small scale of minimal functionalities or large scales, e.g., in the case of ML, it can be one step of data analysis or the entire data preparation. The scale of modularity and abstraction influences the amount of effort needed. Small-scale modularity allows programmers to use the individual parts of the model but requires a more specific framework to combine the code. In comparison, modularity on a larger scale allows for more flexibility in the framework but less reuse of specific code. Consequently, there is a time trade-off between code reuse and time spent on modularity. Registries often use meta-coding principles, the *meta* composability, to register and call modules. This is to limit computational overhead, but registries are still susceptible to computational overhead in frequent calls when the assignment is often overwritten or is not a pointer assignment.

**Reusability.** Starting from scratch or reconstructing previously written code is an inefficient way to build on previous research. Reusable code instead allows researchers to bypass this initial stage and directly build upon previous research.

When previous work contains an implemented registration system, new work can automatically use the entire system and easily adapt or exchange any registered module [24]. Furthermore, previously registered modules can be called from a new system. In contrast, previous research can also be refactored to support registration, either by modularizing the specific functionalities of interest or by modularizing and registering the section of the execution that requires change. Consequently, registered modules are easily reused in new research and new research can easily build upon old implementations. Both forward-implemented and backward-implemented registries bypass initial avoidable time commitments and allow researchers to focus on the new research immediately.

**Interchangeability.** Registries for models also help to streamline the process to benchmark various methods. Similarly, a registry of datasets aids the process of applying the same method to varying benchmarks. Easily switching between implementations can increase experimentation speed in writing code connected with new research. Modularity allows researchers to experiment more easily with various options for code applications. Registration, in turn, further allows researchers to variate, update, adapt, and modify modules because registered modules are inherently modular. This automatically forces code to adhere to the five R's of published work [5], as discussed in the introduction. Depending on the abstraction level of the modules, other researchers can variate and use the coding framework, as well as easily replace and reuse parts of it. Furthermore, it increases code longevity as outdated code can be easily updated. Similarly, due to increased reuse by other researchers, citations and longevity increase as the original author is no longer the only one with a vested interest in the original code.

**Tools.** Registration of modules inside the toolkit can allow users to retrieve modules and generate new applications following tool specifications. The latter decreases the threshold for new functionalities to be suggested to the main tool. Moreover, registries allow researchers to build functionalities in private repositories using the framework set out by the toolkit and then easily upload those, once the original work has been published. As previously stated, some available tools already provide versions of the registration feature [24,47,49]. However, their scope is mostly limited to the registration of ML models. The main advantage of our proposed generic registration, which is missing in previous implementations, is the flexibility to make specific design choices.

**Clarity.** Research collaborations in multidisciplinary fields, such as drug discovery, rely on integrating code from various sources and contributors, thus requiring deliberate forethought and coordination. As such, a standardized workflow is of particular importance and written code should be modular to avoid code instability when working on different interdependent code sections. The registration system creates a clear structure to use the registered modules, thus allowing researchers or project coordinators to standardize their workflows and call functionalities through registries where code can be varied. Consequently, registration increases the efficiency of research collaborations.

**Stability.** The stability of a codebase can be affected when new modules are added. As new functionality is introduced, the potential for interactions and conflicts with existing code increases. This can lead to defects and unexpected behavior. Additionally, adding new modules to a codebase can increase its complexity and make it more difficult to maintain and understand, which can lead to issues in the long term. To mitigate these risks, it is important to have a thorough testing process in place before new modules are added and to thoroughly review and test the entire codebase after the new modules are integrated. A registry can be used to increase the stability of code through specified points to integrate new modules as well as the ability to introduce post-registration checks. These post-checks can then be used to enforce adaptation of external code to the existing framework, such as passing a set of tests or adherence to meta information, such as versioning and factory patterns. As an example, registries can be set up to handle different versions of the same module such that modules are registered together with their version. This way users can more easily track the influence of changing modules. Additionally, the modularity of new functionality ensures minimal impact on the existing codebase.

### 4.4   Codebases for Efficient Coding in Research

Codebases focus more on the software development process. As such, a codebase should particularly ease continuous development, aid code stability, and allow for incremental addition of modules. Due to the multidisciplinary field, ML research focused on drug discovery uses multiple external tools. The usage of tools ranges from the curation and processing of data to the general setup and deployment of models. While the ML pipelines are primarily created for individual projects separately, the developed functionalities can often be helpful for other projects or researchers in the same field. Functionalities from individual projects are often presented with an irregularity in the level of modularity and, therefore, accessibility and usability.

For collaborations or big projects, a choice is made between keeping multiple single-application repositories or creating a bigger shared codebase. There is an open discussion on the advantages and disadvantages of codebases over multiple single-project repositories [20]. Codebases represent an opportunity for collaborators to standardize their workflows as well as share and reuse code with particular functionalities. In general, more standardized workflows allow for more specific coding criteria. Modular code or modular tools are parts of the workflow that can be easily updated with or interchanged for applications with similar functionality. More modular codebases increase collaborator contribution [4] and allow for more use of external tools within its framework. By using registries when designing a codebase, active choices can be made to promote modularity in the project. The benefits of choosing a high level of modularity include code longevity, reuse, and increased potential for collaboration and research speed. Therefore, using a codebase can widen the scope of single, multidisciplinary research projects, making them more modular and reusable for other projects or researchers.

On the other hand, there are disadvantages to coding in a codebase environment: 1) Even though the level of modularity is flexible, this level should be static during research development to prevent unnecessary overhead. It can otherwise be costly to uphold the modularity and maintain backward compatibility. 2) Codebases can restrict reproducibility. Reproducibility requires specific versioning, which can be more fluid in the continuous development within codebases. 3) Codebases can introduce irrelevant functionalities that obscure crucial functionalities in open-source publications. 4) Although coding in a codebase can speed up research long-term, setting up research using modular code is more time-consuming.

However, we advocate for a more general perspective on codebases where any published code can be considered a codebase. This interpretation of published code is less dependent on actual features of codebases, as most code in research often does not require active design, nor intricate design patterns to function properly. Instead, we view the act of writing code as actively assuming others will reuse parts of the code, which will then promote the idea of modularity and reusability, including the use of registries where warranted. This view is more flexible in its application and gives the scientist a base from which to work.

## 4.5 Future Work

The capabilities of registries can be further explored. For example, future research can further investigate how registration aids reproducibility in practice through experiments or surveys. As previously mentioned, continuous development and non-contributing code can impede reproducibility and clarity. Similarly to the accreditation system, versioning, and optional factory patterns, other modules could be attached to registered modules in the registry to aid stability and reproducibility. One can even imagine a generative functionality of the registry to produce a single repository of only the necessary modules from a codebase for a specific application.

A common issue that can occur when combining modules from different sources is that versions can be incompatible. Registries, as proposed here, would not immediately solve problems with dependency conflicts but one can imagine an extension where the registration of modules is accompanied by dependency requirements that are automatically checked and installed upon execution. This in turn will not deal with situations where different parts of the code use different versions of third-party dependencies. However, it would allow for the registration of modules that depend on conflicting versions into the same registry where only one at a time is used during execution.

Furthermore, future research might analyze how their effectiveness depends on using a generic registry. While higher usage promotes modularity, it also removes a level of visibility. An analysis of the overall usage and the usage focused on specific groups of modules will give a better understanding of the best practices of registries. Likewise, one can investigate the trade-off between code visibility and coding speed as noted by Jaspan et al. [20]. Ultimately, collaborating researchers should investigate whether using registries and codebases instead of multiple repositories is advantageous for their research goals and try to design their code to best suit their needs.

## 5   Conclusions

To conclude, we highlight the importance and promise of the registry design pattern, especially in the field of ML development for drug discovery. Registries can promote code reuse through their modular nature. Modularity is the independence of a module from the rest of the code and is crucial for reuse. A registry also promotes other important coding practices and includes the possibility to easily switch between custom functionalities and functionalities from open-source tools. We introduce a method to flexibly register objects and add additional functionalities such as accreditation and versioning. Additionally, we outline the advantages and considerations of registries and stress that registries clarify the usually concealed abstraction and composability of a system. Finally, registries promote clarity, experimentation speed, good coding practices, and code reuse.

**Acknowledgements.** We thank our colleagues and reviewers for their valuable feedback. Additionally, the authors thank BioRender for the publication of the figures generated with BioRender authorized under the subscription plan of Peter Hartog. This study was partially funded by the European Union's Horizon 2020 research and innovation programme under the Marie Skłodowska-Curie Actions grant agreement "Advanced machine learning for Innovative Drug Discovery (AIDD)" No. 956832.

**Disclosure of Interests.** The authors have no competing interests to declare that are relevant to the content of this article.

## References

1. Abadi, M., et al.: TensorFlow: a system for large-scale machine learning. In: 12th USENIX Symposium on Operating Systems Design and Implementation (OSDI 16), pp. 265–283. USENIX Association (2016)
2. Almadi, S.H., Hooshyar, D., Ahmad, R.B.: Bad smells of gang of four design patterns: a decade systematic literature review. Sustainability **13**(18), 10256 (2021)
3. Baker, M.: Reproducibility crisis. Nature **533**(26), 353–66 (2016)
4. Baldwin, C.Y., Clark, K.B.: The architecture of participation: does code architecture mitigate free riding in the open source development model? Manage. Sci. **52**(7), 1116–1127 (2006)
5. Benureau, F.C., Rougier, N.P.: Re-run, repeat, reproduce, reuse, replicate: transforming code into scientific contributions. Front. Neuroinform. **11**, 69 (2018)

6. Cadwallader, L., Hrynaszkiewicz, I.: A survey of researchers' code sharing and code reuse practices, and assessment of interactive notebook prototypes. PeerJ **10**, e13933 (2022)

7. Chen, H., Engkvist, O., Wang, Y., Olivecrona, M., Blaschke, T.: The rise of deep learning in drug discovery. Drug Discov. Today **23**(6), 1241–1250 (2018)

8. Chollet, F., et al.: Keras (2015). https://keras.io

9. da Costa, V.G.T., Fini, E., Nabi, M., Sebe, N., Ricci, E.: solo-learn: a library of self-supervised methods for visual representation learning. J. Mach. Learn. Res. **23**(56), 1–6 (2022)

10. Dallago, C., et al.: Learned embeddings from deep learning to visualize and predict protein sets. Curr. Protoc. **1**(5), e113 (2021)

11. Dara, S., Dhamercherla, S., Jadav, S.S., Babu, C., Ahsan, M.J.: Machine learning in drug discovery: a review. Artif. Intell. Rev. **55**(3), 1947–1999 (2022)

12. Fowler, M., Rice, D., Foemmel, M., Hieatt, E., Mee, R., Stafford, R.: Patterns of Enterprise Application Architecture. Addison-Wesley Professional (2002)

13. Gamma, E., Helm, R., Johnson, R., Vlissides, J.M.: Design Patterns: Elements of Reusable Object-Oriented Software. Pearson Deutschland GmbH, Munich (1995)

14. Goyal, S.: More data science, less engineering: a Netflix original. In: 2020 USENIX Conference on Operational Machine Learning (2020)

15. Howard, J., Gugger, S.: FastAI: a layered API for deep learning. Information **11**(2), 108 (2020)

16. Huang, K., et al.: Therapeutics data commons: machine learning datasets and tasks for drug discovery and development. In: Thirty-fifth Conference on Neural Information Processing Systems Datasets and Benchmarks Track (Round 1) (2021)

17. Hunt, A., Thomas, D.: The Pragmatic Programmer. Addison-Wesley, Boston, United States (1999)

18. Hussain, S., Keung, J., Khan, A.A.: The effect of gang-of-four design patterns usage on design quality attributes. In: 2017 IEEE International Conference on Software Quality, Reliability and Security (QRS), pp. 263–273. IEEE (2017)

19. Hussain, S., Keung, J., Khan, A.A., Bennin, K.E.: Correlation Between the Frequent Use of Gang-of-four Design Patterns and Structural Complexity. In: 2017 24th Asia-Pacific Software Engineering Conference (APSEC), pp. 189–198. IEEE (2017)

20. Jaspan, C., et al.: Advantages and disadvantages of a monolithic repository: a case study at Google. In: Proceedings of the 40th International Conference on Software Engineering: Software Engineering in Practice, pp. 225–234 (2018)

21. Landrum, G.: RDKit: Open-Source Cheminformatics (2006). https://doi.org/10.5281/zenodo.6961488, http://www.rdkit.org

22. Lhoest, Q., et al.: Datasets: a community library for natural language processing. In: Proceedings of the 2021 Conference on Empirical Methods in Natural Language Processing: System Demonstrations, pp. 175–184. Association for Computational Linguistics (2021)

23. Liskov, B.H., Wing, J.M.: A behavioral notion of subtyping. ACM Trans. Program. Lang. Syst. **16**(6), 1811–1841 (1994)

24. Manica, M., et al.: Accelerating material design with the generative toolkit for scientific discovery. NPJ Comput. Mater. **9**(1), 69 (2023)

25. Martin, R.C.: The dependency inversion principle. C++ Report **8**(6), 61–66 (1996)

26. Martin, R.C.: Design principles and design patterns. Object Mentor **1**(34), 597 (2000)

27. Mary, H., et al.: Datamol: molecular manipulation made easy (2022). https://doi.org/10.5281/zenodo.6856321, https://datamol.io/

28. Minnich, A.J., et al.: AMPL: a data-driven modeling pipeline for drug discovery. J. Chem. Inf. Model. **60**(4), 1955–1968 (2020)
29. Gnu general public license, version 3. https://opensource.org/licenses/MIT. Accessed 17 January 2022
30. Mo, R., Cai, Y., Kazman, R., Xiao, L., Feng, Q.: Decoupling level: a new metric for architectural maintenance complexity. In: 2016 IEEE/ACM 38th International Conference on Software Engineering (ICSE), pp. 499–510. IEEE (2016)
31. O'Boyle, N.M., Banck, M., James, C.A., Morley, C., Vandermeersch, T., Hutchison, G.R.: Open Babel: an open chemical toolbox. J. Cheminf. **3**(1), 1–14 (2011)
32. Paszke, A., et al.: PyTorch: an imperative style, high-performance deep learning library. In: Advances in Neural Information Processing Systems, vol. 32 (2019)
33. Pressman, R.S.: Software Engineering: A Practitioner's Approach. Palgrave Macmillan, Gurgaon, India (2005)
34. Ramsundar, B., Eastman, P., Walters, P., Pande, V.: Deep Learning for the Life Sciences: Applying Deep Learning to Genomics, Microscopy, Drug Discovery, and More. O'Reilly Media, Sebastopol (2019)
35. Sarjoughian, H.S.: Model composability. In: Proceedings of the 2006 Winter Simulation Conference, pp. 149–158. IEEE (2006)
36. Sieb, C., Meinl, T., Berthold, M.R.: Parallel and distributed data pipelining with KNIME. Mediterr. J. Comput. Netw. **3**(2), 43–51 (2007)
37. Stålring, J.C., Carlsson, L.A., Almeida, P., Boyer, S.: AZOrange - high performance open source machine learning for QSAR modeling in a graphical programming environment. J. Cheminf. **3**(1), 1–10 (2011)
38. Steinbeck, C., Han, Y., Kuhn, S., Horlacher, O., Luttmann, E., Willighagen, E.: The Chemistry Development Kit (CDK): an open-source java library for chemo- and bioinformatics. J. Chem. Inf. Comput. Sci. **43**(2), 493–500 (2003)
39. Sullivan, K.J., Griswold, W.G., Cai, Y., Hallen, B.: The structure and value of modularity in software design. ACM SIGSOFT Softw. Eng. Notes **26**(5), 99–108 (2001)
40. Sushko, I., et al.: Online Chemical Modeling Environment (OCHEM): web platform for data storage, model development and publishing of chemical information. J. Comput. Aided Mol. Des. **25**(6), 533–554 (2011)
41. Tomar, V., Mazumder, M., Chandra, R., Yang, J., Sakharkar, M.K.: Small molecule drug design. In: Ranganathan, S., Gribskov, M., Nakai, K., Schönbach, C. (eds.) Encyclopedia of Bioinformatics and Computational Biology, pp. 741–760. Academic Press, Oxford (2019)
42. Vamathevan, J., et al.: Applications of machine learning in drug discovery and development. Nat. Rev. Drug Discov. **18**(6), 463–477 (2019)
43. William, F.: PyTorch Lightning (2019). https://doi.org/10.5281/zenodo.3828935, https://www.pytorchlightning.ai
44. Wójcikowski, M., Zielenkiewicz, P., Siedlecki, P.: Open Drug Discovery Toolkit (ODDT): a new open-source player in the drug discovery field. J. Cheminf. **7**(1), 1–6 (2015)
45. Wolf, T., et al.: Transformers: state-of-the-art natural language processing. In: Proceedings of the 2020 Conference on Empirical Methods in Natural Language Processing: System Demonstrations, pp. 38–45. Association for Computational Linguistics (2020)
46. Xiang, Y., Pan, W., Jiang, H., Zhu, Y., Li, H.: Measuring software modularity based on software networks. Entropy **21**(4), 344 (2019)
47. Zaharia, M., et al.: Accelerating the machine learning lifecycle with MLflow. IEEE Data Eng. Bull. **41**(4), 39–45 (2018)

48. Zhang, C., Budgen, D.: What do we know about the effectiveness of software design patterns? IEEE Trans. Softw. Eng. **38**(5), 1213–1231 (2011)
49. Zhu, Z., et al.: TorchDrug: A powerful and flexible machine learning platform for drug discovery. arXiv preprint arXiv:2202.08320 (2022)

**Open Access** This chapter is licensed under the terms of the Creative Commons Attribution 4.0 International License (http://creativecommons.org/licenses/by/4.0/), which permits use, sharing, adaptation, distribution and reproduction in any medium or format, as long as you give appropriate credit to the original author(s) and the source, provide a link to the Creative Commons license and indicate if changes were made.

The images or other third party material in this chapter are included in the chapter's Creative Commons license, unless indicated otherwise in a credit line to the material. If material is not included in the chapter's Creative Commons license and your intended use is not permitted by statutory regulation or exceeds the permitted use, you will need to obtain permission directly from the copyright holder.

# Artificial Intelligence Methods for Evaluating Mitochondrial Dysfunction: Exploring Various Chemical Notations Suitable for Neural Language Processing Models

Edoardo Luca Viganò[1](✉)(iD), Erika Colombo[1](iD), Davide Ballabio[2](iD), and Alessandra Roncaglioni[1](iD)

[1] Laboratory of Environmental Toxicology and Chemistry, Department of Environmental Health Sciences, Istituto di Ricerche Farmacologiche Mario Negri IRCSS, Milan, Italy
Edoardo.vigano@marionegri.it
[2] Milano Chemometrics and QSAR Research Group, Department of Earth and Environmental Sciences, University of Milano-Bicocca, 20126 Milan, Italy

**Abstract.** In recent years, the integration of Artificial Intelligence and Machine Learning methods, such as Neural Language Processing (NLP), with biochemical and biomedical research has revolutionized the field of toxicology defining a profound advancement in our understanding of the toxicological effects of diverse chemical compounds on biological systems.

Among various toxic effects, mitochondrial dysfunction has emerged as a crucial endpoint due to its role in various diseases related to the liver, heart brain, and more in general related to different physiological processes. Indeed, mitochondria are indispensable organelles in cells that serve as the primary hub for energy production, and they are responsible for critical functions in cell metabolism, signaling, and cellular demise. Traditional methods for assessing chemical hazards and their impact on mitochondrial function have relied heavily on experimental assays and animal studies, which are often time-consuming, resource-intensive, and limited in scalability. To overcome these limitations, in silico methods have emerged as indispensable tools in toxicological research to reduce the need for traditional in vivo testing and saving valuable resources in terms of time and money.

This study utilized NLP models to explore diverse chemical notations utilized to encode chemical information such as Simplified Molecular Input Line Entry System (SMILES), DeepSMILES and Self-Referencing Embedded Strings (SELFIES), with the aim of evaluating toxic interactions between chemicals and specific biological targets, achieving high predictivity performance.

**Keywords:** Mitochondrial Dysfunction · Artificial Intelligence · Neural Language Processing · Toxicology

© The Author(s) 2025
D.-A. Clevert et al. (Eds.): AIDD 2024, LNCS 14894, pp. 116–131, 2025.
https://doi.org/10.1007/978-3-031-72381-0_10

# 1    Introduction

In recent years, the integration of machine learning (ML) and artificial intelligence (AI) with biochemical and biomedical research has opened new avenues for understanding the toxicological effects of various chemicals on biological systems. Among these effects, mitochondrial dysfunction has emerged as a critical area of investigation due to its implications for numerous diseases and physiological processes [4,11]. Indeed, mitochondria play a central role in cell metabolism, serving as fundamental components in energy production, metabolism, and cellular signaling. Dysfunction in these organelles can lead to a wide range of health issues, including neurodegenerative diseases, and metabolic, liver, and cardiac disorders [8,16]. Remarkably, chemical compounds have been shown to disrupt mitochondrial function through various mechanisms, such as inducing oxidative stress, disrupting the electron transport chain, or inhibiting other crucial mitochondrial processes which can also result in adverse effects like chemicals-induced liver injury [8,16,25]. These reasons highlight the importance of early identification of potential mitochondrial toxic compounds during the drug development process to mitigate the risk of adverse reactions and toxic effects.

Currently, the assessment of mitochondrial toxicity has heavily relied on in vitro assays measuring specific endpoints, such as alterations in membrane potential or inhibition of the respiratory chain [13]. However, these assays often require significant time and resource investments and may fail to capture the full spectrum of effects that a compound can have on mitochondrial function. Moreover, they necessitate a priori knowledge of the mechanism of toxicity, which may not always be readily available for newly synthesized compounds.

To overcome these limitations, in silico methods have emerged as indispensable tools in toxicological research and artificial intelligence (AI) and machine learning (ML) are among today's advanced approaches for evaluating chemical hazards. Diverse ML and AI models can be found in literature developed to predict mitochondrial dysfunction using different methodologies based on Quantitative Structure Activity Relationship (QSAR) [5,11,23,27].

Also, more recent advances in high-throughput imaging technologies have enabled the generation of large-scale datasets that capture cellular and organelle morphology in response to chemical perturbations, which can be used to improve models' predictions [15,24]. One of these works tested multiple ways to encode chemical information revealing that one of the most promising approaches to manage mitochondrial dysfunction is using neural language processing (NLP) directly on chemical notation [27]. Based on the evidence provided by the literature, we decided to further explore the prediction efficacy of NLP models using different chemical notations.

These NLP methods offer a data-driven and computationally efficient approach to toxicological screening by predicting chemical toxicity directly from chemical notations. Leveraging advancements in deep learning and neural language processing, these methods hold promise for accelerating drug discovery, environmental risk assessment, and chemical safety evaluation.

NLP models for predicting chemical toxicity from chemical notation often utilize sequence-to-sequence (Seq2Seq) models or variants. The architecture of these networks can process chemical notation directly as text using specific layers for text vectorization and character embedding, which describes each character or segment of the string as a numerical vector. These models, such as Recurrent Neural Networks (RNNs), Long Short-Term Memory (LSTM), or transformers, are designed to handle sequential data and capture dependencies between tokens in the input sequence and the properties.

In this study, we focus on exploring various chemical notations that can serve as input for NLP methods. We evaluated the SMILES (Simplified Molecular Input Line Entry System) that is a widely used chemical notation system that represents the structure of molecules as a linear string of characters, DeepSMILES [17] and Self-Referencing Embedded Strings (SELFIES) [12]. DeepSMILES and SELFIES are adaptations of SMILES designed to address some of the issues that arise when using strings to represent chemicals in machine learning. For instance, DeepSMILES avoids the problem of pairing ring closure symbols by using only a single symbol at the ring closing location, where the symbol indicates the ring size. SELFIES (Self-Referencing Embedded Strings) instead is designed to be a more robust notation since every possible SELFIES string corresponds to a valid molecule. These differences in notation can have significant impacts on the performance of machine learning models trained on these representations.

## 2    Material and Methods

### 2.1    Dataset

We collected data from the ICE database (https://ice.ntp.niehs.nih.gov/ accessed on 15 October 2023), which provides high-quality curated data to support the development and evaluation of new, revised, and alternative methods. Table 1 lists the assays used.

We selected datasets containing specific biological assays that have a critical role in identifying potentially harmful compounds able to interact as stressors for the increase in mitochondrial dysfunction.

The chemicals retrieved are univocally defined by the CAS number. We then retrieved the SMILES from CAS using in-house software (https://github.com/EdoardoVigano/Chemical-Resolver). We represent compounds using SMILES because it is a widely used and standardized notation system for representing chemical structures and molecules, using text strings.

We curated the retrieved SMILES by performing standard SMILES canonization, followed by the removal of structures displaying inconsistencies that might indicate chemical errors. We also excluded stereochemistry and removed salts, concentrating solely on the largest fragments. Any duplicate structures were removed, maintaining only one instance in cases where the experimental values were consistent between duplicates. This approach to SMILES curation is very commonly used [9].

**Table 1.** Assays to evaluate mitochondrial dysfunction.

| ASSAYS to Increase in Mitochondrial Dysfunction |
| --- |
| APR HepG2 MitoMass 24 h dn |
| APR HepG2 MitoMass 24 h up |
| APR HepG2 MitoMass 72 h dn |
| APR HepG2 MitoMass 72 h up |
| APR HepG2 MitoMembPot 24 h dn |
| APR HepG2 MitoMembPot 24 h up |
| APR HepG2 MitoMembPot 72 h dn |
| APR HepG2 MitoMembPot 72 h up |
| ATG XTT Cytotoxicity up |
| TOX21 MMP ratio down |
| TOX21 MMP ratio up |
| TOX21 MMP rhodamine |

The data are labeled as active or inactive for classification modeling purposes. We define activity as follows: a chemical is considered active if it shows a hit call label as active in at least one of the selected assays for that specific biological target; otherwise, it is labeled as inactive. The individual assay label was available in the files downloaded from the ICE platform.

The raw data provided by a vendor or laboratory underwent processing, indexing, transformation, and normalization using standardized methods. Subsequently, the concentration-response data are subjected to modeling through three selected models (constant, Hill, and gain-loss). If any models fit sufficiently, the chemical assay pair is considered 'active' (hit call = active); otherwise, the final hit call is 'inactive'. Characteristics of the data collected, and class proportions are shown in Table 2.

**Table 2.** Summary of data for mitochondrial dysfunctions with information about number of compounds, and percentages of active and inactive.

| Name | $N$ | Active | Inactive | Active % | Inactive % | $N$ of Assays |
| --- | --- | --- | --- | --- | --- | --- |
| Mitochondrial Dysfunction | 5004 | 1147 | 3857 | 23 | 77 | 12 |

The datasets is unbalanced, and the statistics on dataset composition are reported in Table 2. That information is important to consider applying a method for oversampling the minority class and selecting the right metrics to assess the model's performance.

## 2.2    Chemical Notations and Data Augmentation

Chemical notation refers to the standardized symbols and conventions used to represent chemical compounds and other chemical phenomena. In this work, we explore three different chemical notations: SMILES, SELFIES, and DeepSMILES (Fig. 1). All these chemical notations represent the chemical as a list of text characters suitable for NLP models.

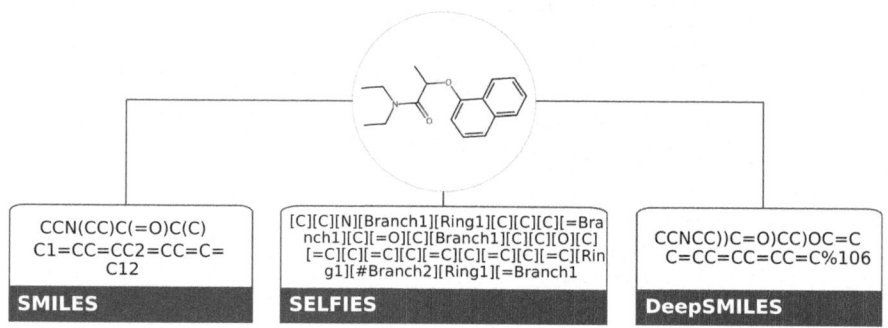

**Fig. 1.** Chemical notations explored as input in NLP models.

To encode chemical information from chemical notations various approaches exist, such as the conventional methods typically employed in QSAR as molecular descriptors (MDs) [7], or different types of fingerprints like Morgan fingerprint and Molecular ACCess System (MACCS). Other possible methods involve graph [6], latent representations generated by encoder-decoder architecture [10, 28] and character embeddings commonly used in neural language processing. In our work we focused on the latter since this approach is computation efficient and very promising.

Character/word embedding is applied directly to the chemical notation, as a string of characters, without any other considerations or intermediate steps. Tokenization involves breaking the text into smaller units known as tokens, that can be a single character or group of characters. We adopted two different methods to tokenize the string of characters: one is the atomwise tokenizer, and the other is the kmer tokenizer with ngram parameter of 4 [14, 22]. Kmer with ngram of 4 meaning the length of each segment contains 4 characters.

To determine the maximum length of tokenized molecules, we calculated the number of tokens for each chemical and then utilized the value corresponding to the 95th percentile. In Fig. 2 is reported an example of different tokenizer methods to break the chemical notation for instance on DeepSMILES.

The tokenized chemical notation can be used as input in an embedding layer where each token is represented by a numerical vector in the network. The model learns this embedding during training, which is one of the most significant advantages compared to classical techniques such as Molecular Descriptors used in QSAR, where chemicals are described in a predefined manner.

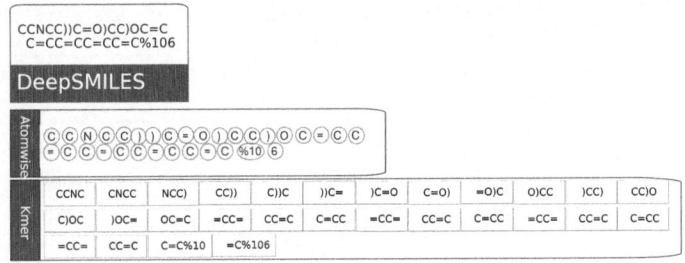

**Fig. 2.** Comparison of the results of the two different tokenizers used, the atomwise and the kmer.

With the aim of further improving the performance of the NLP models, we employed a data augmentation process. Data augmentation is an essential technique for enhancing the diversity and size of training data in machine learning, especially when the data is limited. We utilized the SMILES enumeration approach [4]. Here, the fact that multiple SMILES representations correspond to the same molecule is leveraged as a technique to expand the dataset. With this method, we were able to increase the dataset by a factor of 10.

## 2.3   Data Pre-processing

It is common practice to split the original dataset into training, validation, and testing sets. We initially split the data into 90% for training and 10% for hold-out test set. Then, within the training data, we further split it into 90-10% to generate the validation set. In both splits, we maintained the ratio between the toxic and non-toxic labels as in the original dataset. The results are reported in Table 3.

The data preparation for AI architectures must consider the computational cost and optimize the training time to create the best conditions for model convergence and stability during training. One way to achieve this is by normalizing activations within each mini-batch, which can reduce the likelihood of overfitting.

In the case of our NLP architecture, the dataset was divided into mini-batches of 32, and we employed TensorFlow for data prefetching. This technique is used to enhance the training performance of deep learning models by overlapping the data loading and model training phases. The goal of data prefetching is to minimize the idle time of the GPU or CPU during training and mitigate the impact of data loading latencies on overall training speed. By keeping the computational units, such as GPUs or CPUs, fully utilized, data prefetching achieves faster training times and more efficient model convergence.

**Table 3.** Summary of Mitochondrial Dysfunction data after preprocessing phase.

| Mitocondrial Dysfunction | All dataset | Training set | Validation set | Holdout set |
|---|---|---|---|---|
| All | 5004 | 4052 | 451 | 501 |
| Active | 1147 | 929 | 103 | 115 |
| Inactive | 3857 | 3123 | 348 | 386 |

### 2.4   Applicability Domain

To calculate the applicability domains (AD) of our models, we used the Applicability Domain Toolbox (Milano Chemometrics and QSAR Research Group) [1]. This tool implements a set of AD approaches based on several strategies, such as range-based methods, geometric methods, and distance-based methods.

In this work specifically we selected the 'Bounding box PCA', the 'Leverage', the 'Distance from centroid', and the 'Distance kNN with predefine k to 5' methods to define the AD [19–21]. These methods are used to reach a consensus; therefore, if a chemical in the test set is defined as out of domain by at least three out of four of these algorithms, it is considered potentially to be discarded.

This evaluation was conducted on a dataset where chemicals were encoded using latent representations generated by the SeqToSeq encoder-decoder architecture trained on SMILES notation [28].

### 2.5   Model Architecture

We employed an NLP model with an architecture able to work directly on chemical notations as strings of text. This way, the networks learn the chemical notation grammar and correlations between strings or sequences of characters with biological target interactions.

The network architecture consists of several layers designed to process input data in the form of chemical notation text strings. It begins with the Input Layer, which receives the text strings representing chemical notations. These strings are then converted into numerical vectors by the Text Vectorization Layer, specifically designed for converting SMILES strings. The next step is the Embedding Layer that represents each tokenized character from the chemical notation strings as a dense vector. Convolutional Layers (Conv1D) are then employed to perform convolution operations, capturing local patterns within the chemical notation strings. The results of the convolution serve as input for the Bidirectional LSTM layers, which are designed to capture contextual information from both preceding and succeeding tokens in the chemical notation strings. The Global Max Pooling 1D Layer is then applied to extract the most relevant features from the output of the bidirectional LSTM layers. Subsequently, Dense Layers perform fully connected operations to further process the extracted features and Batch Normalization and Dropout Layers are employed to enhance training stability and prevent overfitting by normalizing and randomly dropping out units during training. Finally, the Output Layer produces the final output of the classification prediction. Together, these layers form a comprehensive network architecture optimized for processing and classifying chemical notation data.

Table 4 provides a more detailed summary of the NLP model architecture used for each chemical notation. Notably, although the architecture remains the same for each of the NLP models developed, to mitigate specific cases of overfitting or poor performance, a grid search was performed on different model parameters. These parameters included the embedding dimension, the number of convolution filters and the number of hidden layers in LSTM.

**Table 4.** Summary of NLP model architecture.

| Layer (type) | Output Shape | Param # |
| --- | --- | --- |
| Input Layer | [(None, 1)] | 0 |
| Text Vectorization | (None, 54) | 0 |
| Embedding | (None, 54, 128) | 2091776 |
| Conv1D | (None, 54, 128) | 328192 |
| Conv1D | (None, 54, 64) | 655616 |
| Bidirectional | (None, 54, 64) | 394240 |
| Bidirectional | (None, 54, 32) | 123648 |
| GlobalMaxPooling1D | (None, 128) | 0 |
| Dense | (None, 128) | 16512 |
| Batch Normalization | (None, 128) | 512 |
| Dropout | (None, 128) | 0 |
| Dense | (None, 128) | 1056 |
| Batch Normalization | (None, 128) | 128 |
| Dropout | (None, 128) | 0 |
| Dense | (None, 64) | 528 |
| Dense | (None, 1) | 17 |

Tot. param: 2303081 (8.79 MB)

Trainable params: 2302953 (8.79 MB)

Non-trainable params: 128 (512.00 Byte)

## 2.6   Model Training and Model Validation

One of the most important limitations we encountered in reaching good performance for modeling is the presence of unbalanced classes. This is a very common problem with real data provided by biological assays, and data distribution makes it challenging for a model to learn and predict the minority class effectively. In addition, models operating on imbalanced data can seem very accurate when measured by traditional accuracy metrics, yet they may perform poorly in practice. We employed two different approaches to address these issues. One of them involves weighting the loss higher for the minority class during training, which is a common practice to help overcome the imbalance problem. The other method we adopted was to initialize the final layer weights considering the unbalanced proportion of our dataset, rather than using the common method of random initialization. Indeed, considering the class ratio, it is possible to set the bias on the logits such that the network predicts a predefined probability at initialization. Setting these correctly will speed up convergence and account for the unbalanced dataset at the same time, in this way during the first few iterations network is not limited to essentially just learning the bias.

After the pre-processing phase, the data can be used to train and validate models with internal and holdout validation. The models are trained on the training dataset for multiple epochs. During each epoch, input data are processed through the model and model parameters are updated based on the prediction errors evaluated on the validation set. This iterative process enables to track the history of your selected optimization metric as it converges to a minimum over epochs. To further optimize computational cost and improve model performance, prevent overfitting, and make the training process more efficient, we monitored the validation loss. In cases where during training the model stops to improving or reaches a plateau for a certain number of epochs (patience), the learning rate is reduced by a factor of 0.1. We decided to reduce the learning rate to help the model navigate closer to the optimal point in the parameter space, hopefully finding a better minimum. In cases where, even after the reduction of the learning rate, the model didn't improve for three epochs, we stopped the training phase, assuming the model reached convergence.

For the holdout validation, we incorporated diverse metrics to show a fuller assessment of the model's predictive abilities, so we could evaluate more deeply the model's ability to generalize the knowledge gained from the training set. The metrics selected to evaluate the model's performance on holdout tests are balance accuracy, precision, sensitivity, specificity, Matthews correlation coefficient (MCC), and F1-score.

## 2.7   Software

All models and architecture implementations were performed with Python packages. Python 3.9.16, RDKit (version 2023.03.1), scikit-learn 1.2.2, SciPy 1.8.1, pandas 1.5.3, matplotlib 3.7.1, and deepchem 2.7.1, TensorFlow 2.12.0 and Keras 2.12.0 were used to create architecture for deep learning models as NLP methods.

# 3   Results

We encoded the chemical information using CDDD descriptors and defined whether the chemicals were outside or inside the AD using a consensus approach based on 'Bounding box PCA', 'Leverage', 'Distance from centroid', and 'Distance kNN with a predefined k of 5'.

There were 4 chemicals considered outside of the domain, and they were discarded from the test set (CAS: 147536-97-8, 81-55-0, 40220-08-4, 25999-20-6).

The performance of the NLP models is reported for each of the tested chemical notations. Each of these notations was tokenized in two different ways: using the atomwise tokenizer and the kmer tokenizer, which fragment the strings of characters differently. For a better comparisons between our methodologies and machine learning, we added in the Table 5 the results from literature regarding baseline models to predict mitochondrial dysfunction [11,15,27].

The overall results are presented in Table 5.

**Table 5.** Summary of NLP model results obtained. In green is reported the results obtained for SMILES notation with different tokenizer and with or without data augmentation (AUG). In yellow are reported the results for the DeepSMILES notation and in blue the ones related to the SELFIES notation. The red values represent the values of the best models overall. In the last rows in grey, the results obtained in other works using machine learning methods built on conventional molecular descriptors are reported.

| Notation | BA | Prec | Sens | Spec | MCC | F1-Score | Validation F1-Score | Tokenizer | AUG |
|---|---|---|---|---|---|---|---|---|---|
| SMILES atomwise | 0.766 | 0.660 | 0.625 | 0.906 | 0.542 | 0.642 | 0.744 | atomwise | no |
| AUG SMILES atomwise | 0.810 | 0.552 | 0.812 | 0.808 | 0.550 | 0.657 | 0.803 | atomwise | yes |
| SMILES kmer | 0.747 | 0.536 | 0.661 | 0.834 | 0.461 | 0.592 | 0.857 | kmer | no |
| AUG SMILES kmer | 0.761 | 0.717 | 0.589 | 0.932 | 0.561 | 0.647 | 0.613 | kmer | yes |
| DeepSMILES atomwise | 0.785 | 0.557 | 0.741 | 0.829 | 0.519 | 0.636 | 0.731 | atomwise | no |
| AUG DeepSMILES atomwise | 0.764 | 0.503 | 0.741 | 0.787 | 0.469 | 0.599 | 0.745 | atomwise | yes |
| DeepSMILES kmer | 0.764 | 0.503 | 0.741 | 0.787 | 0.469 | 0.599 | 0.822 | kmer | no |
| AUG DeepSMILES kmer | 0.742 | 0.503 | 0.679 | 0.805 | 0.439 | 0.578 | 0.757 | kmer | yes |
| **SELFIES atomwise** | **0.775** | **0.707** | **0.625** | **0.924** | **0.575** | **0.664** | **0.688** | **atomwise** | **no** |
| AUG SELFIES atomwise | 0.749 | 0.605 | 0.616 | 0.883 | 0.495 | 0.611 | 0.887 | atomwise | yes |
| SELFIES kmer | 0.715 | 0.438 | 0.688 | 0.742 | 0.375 | 0.535 | 0.812 | kmer | no |
| AUG SELFIES kmer | 0.738 | 0.551 | 0.625 | 0.851 | 0.456 | 0.586 | 0.613 | kmer | yes |
| **ML from literature** | | Holdout | Set | | | | Validation | Descriptors | |
| Gradient Boosting [11] | 0.708 | 0.573 | 0.467 | 0.948 | - | - | - | Atom Pair FP | - |
| Random Forest [11] | 0.743 | 0.279 | 0.793 | 0.692 | - | - | - | RDKit mol.desc. | - |
| Neural Network [15] | - | 0.45 | 0.68 | 0.88 | 0.48 | 0.54 | 0.62 | CDDD | - |
| Extreme GB [27] | 0.742 | 0.650 | 0.600 | 0.883 | 0.485 | 0.602 | 0.912 | CDDD | SMOTE |

The table presents the results obtained for each of the chemical notations under evaluation. Two different tokenizers were tested for each notation, and the results are provided for both augmented and non-augmented datasets. Based on these results, various comparisons between the methods were conducted. Specifically, our aim was to first determine which method achieved the highest performance overall in the holdout test set. Secondly, we specifically focused on comparing the different notations, tokenizer methods, and/or data augmentation approaches. The purpose of these comparisons is to explore how these various approaches can affect the models' performance.

The metrics we prioritized for comparison were primarily the F1-Score, MCC, and Balanced Accuracy, considering the unbalanced nature of the dataset. Based on these metrics, the model named *SELFIES atomwise* emerged as the top performer overall, outperforming the others in two out of three prioritized metrics achieving the values for Balanced Accuracy of 0.775, an MCC of 0.575, an F1-Score of 0.664, and a considerable precision value of 0.707. However, further consideration must be given, since the sensitivity metric for this specific model is not optimal. Indeed, it is common practice in toxicology field to consider the model's capability to recognize the positive compounds more important than recognize the non-active ones. Although precision has a very good value for this model, it's possible that is not the best to identify the positive compounds and then we thought to highlight also the model called *AUG SMILES atomwise* as a promising candidate as the best model overall. The *AUG SMILES atomwise* model utilizes SMILES as input for the network and employs the atomwise segmentation method for tokenization. Also, the dataset was extended using the data-augmentation techniques. This model achieved a Balanced Accuracy of 0.810, an MCC of 0.550, and an F1-Score of 0.657, and a considerable sensitivity value of 0.802.

Since the aim of our work was to explore different chemical notations and tokenizers to find the most suitable for NLP methodologies to predict a specific toxicological endpoint, we didn't define a priori which model behavior would be preferred (e.g. high precision or high specificity). For this reason, we are searching for the most robust and general model without considering the final aims of our approaches in detail. The prediction model's capability was evaluated by focusing mainly on the performance of the holdout test set rather than the validation set, as the latter is considered during the training process to adjust the learning rate.

Remarkably is the general comparison between the two different tokenizers. The atomwise tokenizer reaches higher performance in each of the chemical notations explored. These findings indicate that breaking down any chemical notations into single characters and using them as tokens in NLP could enhance the model's performance and generalization capabilities in prediction compared to the segmentation provided by the kmer tokenization.

To facilitate a better comparison of the diverse methods used to tokenize the chemical notations, a bar plot is provided in Fig. 3.

In almost all metrics, atomwise tokenization performs slightly better across almost all chemical notations, except for SMILES. One of the main reasons could be related to the fact that the kmer tokenizer increases the vocabulary size significantly given the available data in compared to atomwise tokenizer. Indeed, with atomwise tokenization, we obtained a vocabulary of about 40 to 50 elements depending on the chemical notation, while for the kmer tokenizer, the vocabulary dimensions are between 4000 and 5000. Models with larger vocabularies must overcome some challenges. For instance, larger vocabulary of course may contribute to improved generalization by enabling the model to capture a more diverse pattern, however, there's a potential downside where the dataset is limited. Indeed the model might inadvertently memorize rare or irrelevant pat-

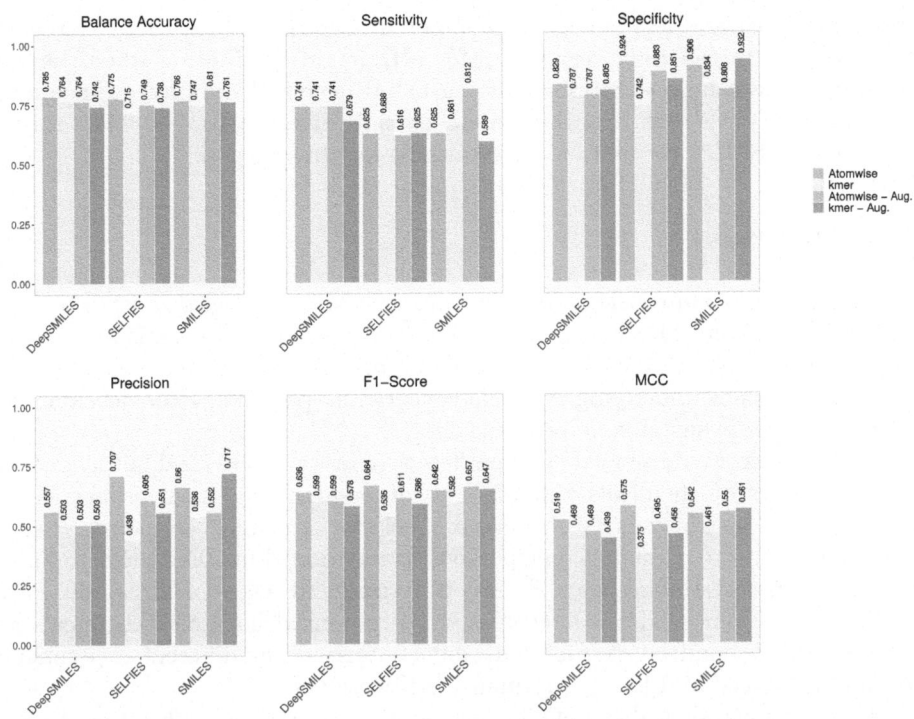

**Fig. 3.** Comparison of the NLP model results across different metrics. The images depict the model performance for all chemical notations and tokenizer methods, highlighting the effect of data augmentation. Models trained on the original dataset with atomwise tokenizer are shown in green. Models trained on the original dataset with kmer tokenizer are shown in yellow. Models trained on the augmented dataset with atomwise tokenizer are shown in lilac. Models trained on the augmented dataset with kmer tokenizer are shown in orange. (Color figure online)

terns present in the expanded vocabulary and this could compromise its ability to perform well on unseen data, thus undermining its generalization capabilities.

Regarding the performance improvement due to data augmentation methods, it seems to have a significant impact only on the SMILES notation. There was a F1-score increase of 2.3% for the atomwise tokenizer and 9.3% for the kmer. The comparison between model performance trained on the original dataset or the augmented ones is also reported in Fig. 3.

Notably, only the NLP models using SMILES notation as input achieved an overall performance improvement with both tokenization methods employed, while SELFIES and DeepSMILES did not show any clear improvement after augmentation. Several factors can explain these results, starting with the fact that the data augmentation approach used in this study was specifically designed for SMILES and not for other chemical notations. SELFIES and DeepSMILES are more robust and less ambiguous compared to SMILES, but these character-istics, which are generally very useful for ML, might be limiting when the aim is to apply data augmentation to string-based chemical notations. Since aug-mentation involves rewriting the same chemicals in various ways, a less robust notation like SMILES could be more suitable for this purpose.

## 4  Conclusion

The use of in silico approaches in New Approach Methodologies-assisted toxicol-ogy, such as AI and ML models, to predict hazards is progressing fast, with the potential to transform the toxicology field by providing a greater understanding of the mechanisms underlying chemical toxicity and permitting the development of safer and more sustainable products.

In this work, we developed NLP models to assess the potential mitochondrial toxicity effects of chemicals belonging to different classes such as pesticides, drugs, and industrializers. We selected precise well-defined endpoints that is fundamental in the regulatory perspective, as mentioned in OECD Environment Health and Safety Publications, Series on Testing and Assessment No. 69, Paris 2007 [2]. With our models, we evaluated the potential hazard of chemicals for increased mitochondrial dysfunction, and we carried out different experiments to build advanced and high-performance AI networks.

This work is a step ahead of what has already been done with ML and AI to predict mitochondrial toxicity effects. Considering the results obtained by other studies [27], we focused our efforts on developing NLP models since they are the most promising. We explored different chemical notations such as SMILES, SELFIES, and DeepSMILES, and different tokenization considering atomwise or kmer tokenizers. Also we applied methodologies for data augmentation and we tried to overcome the problem of unbalanced datasets using different approaches during network training. Our efforts led us to consider the SELFIES represen-tation with atomwise tokenization as the most robust chemical notation for performing NLP methodologies to predict mitochondrial dysfunction with the currently available high-throughput data. These methods offer great promise since they can learn from various chemical notations and further enhance their ability to predict potential toxicity without incurring high computational costs.

One of the limitations of these methodologies is linked to the fact that increas-ing the complexity of models often sacrifices model transparency. The challenge, therefore, lies in crafting models where each layer of complexity serves a purpose, rendering them both powerful and comprehensible. The 'black box' aspect asso-ciated with advanced AI methods leads the final user to prefer more transpar-ent approaches, potentially sacrificing predictive capacity and performance. The need to increase the transparency of these approaches is often recognized, and indeed, a stronger emphasis on the model's explainability is considered crucial. Various methods, including SHAP and LIME [3,18,29], appear to be promising solutions as well as integrating a self-attention mechanism into the network [26] with the goal of exploring the reasoning behind the models' assessments and providing a chemical or toxicological explanation. In the future, we hope to explore the model's interpretability and explainability in different aspects.

**Acknowledgement.** This work was supported by the European Union's Horizon 2020 research and innovation program (grant # 101037090) (project ALTERNATIVE). The content of this abstract reflects only the author's view, and the Commission is not responsible for any use that may be made of the information it contains.

**Disclosure of Interests.** The authors have no competing interests to declare that are relevant to the content of this article.

# References

1. Applicability Domain Toolbox (for MATLAB). https://michem.unimib.it/down load/matlab-toolboxes/applicability-domain-toolbox-for-matlab/
2. Series on Testing and Assessment: Publications by Number - OECD. https:// www.oecd.org/chemicalsafety/testing/series-testing-assessment-publications-number.htm
3. Belle, V., Papantonis, I.: Principles and practice of explainable machine learning. Front. Big Data **4**, 688969 (2021). https://doi.org/10.3389/fdata.2021.688969
4. Bjerrum, E.: SMILES enumeration as data augmentation for neural network modeling of molecules
5. Bringezu, F., Gómez-Tamayo, J.C., Pastor, M.: Ensemble prediction of mitochondrial toxicity using machine learning technology. Comput. Toxicol. **20**, 10018 (2021).https://doi.org/10.1016/j.comtox.2021.100189
6. Coley, C., et al.: A graph-convolutional neural network model for the prediction of chemical reactivity. Chem. Sci. **10**(2), 370–377 (2018). †electronic supplementary information (ESI) available: additional model and dataset details, results, discussion, and see https://doi.org/10.1039/c8sc04228d
7. Consonni, V., Todeschini, R.: Molecular descriptors. In: Puzyn, T., Leszczynski, J., Cronin, M.T. (eds.) Recent Advances in QSAR Studies: Methods and Applications, pp. 29–102. Springer, Dordrecht (2010). https://doi.org/10.1007/978-1-4020-9783-6_3
8. French, K.J., et al.: Differences in effects on myocardium and mitochondria by angiogenic inhibitors suggest separate mechanisms of cardiotoxicity. Toxicol. Pathol. **38**(5), 691–70 (2010). https://doi.org/10.1177/0192623310373775
9. Gadaleta, D., Lombardo, A., Toma, C., Benfenati, E.: A new semi-automated workflow for chemical data retrieval and quality checking for modeling applications. J. Cheminform. **10**(1), 60 (2018). https://doi.org/10.1186/s13321-018-0315-6
10. Gómez-Bombarelli, R., et al.: Automatic chemical design using a data-driven continuous representation of molecules. ACS Central Sci. **4**(2), 268–276 (2018).https:// doi.org/10.1021/acscentsci.7b00572. American Chemical Society
11. Hemmerich, J., Troger, F., Füzi, B., F Ecker, G.: Using machine learning methods and structural alerts for prediction of mitochondrial toxicity. Molecul. Inform. **39**(5), e2000005 (2020). https://doi.org/10.1002/minf.202000005
12. Krenn, M., et al.: SELFIES and the future of molecular string representations. Patterns **3**(10), 100588 (2022). https://doi.org/10.1016/j.patter.2022.100588

13. Krishna, S., Berridge, B., Kleinstreuer, N.: High-throughput screening to identify chemical cardiotoxic potential. Chem. Res. Toxicol. **34**(2), 566–583 (2021).https://doi.org/10.1021/acs.chemrestox.0c00382

14. Li, X., Fourches, D.: SMILES pair encoding: a data-driven substructure tokenization algorithm for deep learning. J. Chem. Inf. Model. **61**(4), 1560–1569 (2021). https://doi.org/10.1021/acs.jcim.0c01127

15. Garcia de Lomana, M., Marin Zapata, P.A., Montanari, F.: Predicting the mitochondrial toxicity of small molecules: insights from mechanistic assays and cell painting data. Chem. Res. Toxicol. **36**(7), 1107–1120 (2023). https://doi.org/10.1021/acs.chemrestox.3c00086

16. Mihajlovic, M., Vinken, M.: Mitochondria as the target of hepatotoxicity and drug-induced liver injury: molecular mechanisms and detection methods. Int. J. Molecul. Sci. **23**(6), 3315 (2022). https://doi.org/10.3390/ijms23063315

17. O'Boyle, N., Dalke, A.: DeepSMILES: an adaptation of SMILES for use in machine-learning of chemical structures (2018).https://doi.org/10.26434/chemrxiv.7097960.v1

18. Ribeiro, M., Singh, S., Guestrin, C.: "Why Should I Trust You?": Explaining the Predictions of Any Classifier, pp. 97–101 (2016). https://doi.org/10.18653/v1/N16-3020

19. Sahigara, F., Ballabio, D., Todeschini, R., Consonni, V.: Assessing the validity of QSARs for ready biodegradability of chemicals: an applicability domain perspective. Curr. Comput. Aided-Drug Design **10**(2), 137–147 (2017). https://doi.org/10.2174/1573409910666140410110241

20. Sahigara, F., Ballabio, D., Todeschini, R., Consonni, V.: Defining a novel k-nearest neighbours approach to assess the applicability domain of a QSAR model for reliable predictions. J. Cheminform. **5**, 27 (2013). https://doi.org/10.1186/1758-2946-5-27

21. Sahigara, F., Mansouri, K., Ballabio, D., Mauri, A., Consonni, V., Todeschini, R.: Comparison of different approaches to define the applicability domain of QSAR models. Molecules **17**(5), 4791–4810 (2012). https://doi.org/10.3390/molecules17054791

22. Sennrich, R., Haddow, B., Birch, A.: Neural machine translation of rare words with subword units. In: Erk, K., Smith, N.A. (eds.) Proceedings of the 54th Annual Meeting of the Association for Computational Linguistics (Volume 1: Long Papers), pp. 1715–1725 (2016). Association for Computational Linguistics (2016). https://doi.org/10.18653/v1/P16-1162

23. Tang, W., Liu, W., Wang, Z., Hong, H., Chen, J.: Machine learning models on chemical inhibitors of mitochondrial electron transport chain. J. Hazard. Mater. **426**, 128067 (2022). https://doi.org/10.1016/j.jhazmat.2021.128067

24. Tian, G., Harrison, P.J., Sreenivasan, A.P., Puigvert, J.C., Spjuth, O.: Combining molecular and cell painting image data for mechanism of action prediction (2022). pages: 2022.10.04.510834 Section: New Results. https://doi.org/10.1101/2022.10.04.510834

25. Varga, Z.V., Ferdinandy, P., Liaudet, L., Pacher, P.: Drug-induced mitochondrial dysfunction and cardiotoxicity. Am. J. Physiol.-Heart Circul. Physiol. **309**(9), H1453–H1467 (2015).https://doi.org/10.1152/ajpheart.00554.2015

26. Vaswani, A., et al.: Attention is all you need. arXiv preprint arXiv:1706.03762 (2023)

27. Viganò, E.L., Ballabio, D., Roncaglioni, A.: Artificial intelligence and machine learning methods to evaluate cardiotoxicity following the adverse outcome pathway frameworks. Toxics **12**(1), 87 (2024). https://doi.org/10.3390/toxics12010087. Number: 1 Publisher: Multidisciplinary Digital Publishing Institute

28. Winter, R., Montanari, F., Noé, F., Clevert, D.A.: Learning continuous and data-driven molecular descriptors by translating equivalent chemical representations. Chem. Sci. **10**(6), 1692–1701 (2018). † electronic supplementary information (ESI) available: detailed information regarding the final model architecture, hyperparameter grid, results and computation time. See https://doi.org/10.1039/c8sc04175j

29. Štrumbelj, E., Kononenko, I.: Explaining prediction models and individual predictions with feature contributions. Knowl. Inf. Syst. **41**(3), 647–665 (2014). https://doi.org/10.1007/s10115-013-0679-x

**Open Access** This chapter is licensed under the terms of the Creative Commons Attribution 4.0 International License (http://creativecommons.org/licenses/by/4.0/), which permits use, sharing, adaptation, distribution and reproduction in any medium or format, as long as you give appropriate credit to the original author(s) and the source, provide a link to the Creative Commons license and indicate if changes were made.

The images or other third party material in this chapter are included in the chapter's Creative Commons license, unless indicated otherwise in a credit line to the material. If material is not included in the chapter's Creative Commons license and your intended use is not permitted by statutory regulation or exceeds the permitted use, you will need to obtain permission directly from the copyright holder.

# Temporal Evaluation of Uncertainty Quantification Under Distribution Shift

Emma Svensson[1,3](✉)📷, Hannah Rosa Friesacher[2,3]📷, Adam Arany[2]📷,
Lewis Mervin[4]📷, and Ola Engkvist[3,5]📷

[1] ELLIS Unit Linz, Institute for Machine Learning, Johannes Kepler University Linz,
4040 Linz, Austria
svensson@ml.jku.at
[2] ESAT-STADIUS, KU Leuven, 3000 Leuven, Belgium
[3] Molecular AI, Discovery Sciences, R&D, AstraZeneca Gothenburg,
431 83 Mölndal, Sweden
[4] Molecular AI, Discovery Sciences, R&D, AstraZeneca Cambridge,
Cambridge CB2 0AA, UK
[5] Department of Computer Science and Engineering, Chalmers University of
Technology, 412 96 Gothenburg, Sweden

**Abstract.** Uncertainty quantification is emerging as a critical tool in
high-stakes decision-making processes, where trust in automated predictions that lack accuracy and precision can be time-consuming and
costly. In drug discovery, such high-stakes decisions are based on modeling the properties of potential drug compounds on biological assays.
So far, existing uncertainty quantification methods have primarily been
evaluated using public datasets that lack the temporal context necessary
to understand their performance over time. In this work, we address the
pressing need for a comprehensive, large-scale temporal evaluation of
uncertainty quantification methodologies in the context of assay-based
molecular property prediction. Our novel framework benchmarks three
ensemble-based approaches to uncertainty quantification and explores
the effect of adding lower-quality data during training in the form of
censored labels. We investigate the robustness of the predictive performance and the calibration and reliability of predictive uncertainty by
the models as time evolves. Moreover, we explore how the predictive
uncertainty behaves in response to varying degrees of distribution shift.
By doing so, our analysis not only advances the field but also provides
practical implications for real-world pharmaceutical applications.

**Keywords:** uncertainty quantification · temporal evaluation ·
distribution shift · deep learning · drug discovery · molecular property
prediction

## 1   Introduction

Uncertainty quantification enables safer and more reliable deployment of
machine learning models in real-world applications by increasing the confidence

© The Author(s) 2025
D.-A. Clevert et al. (Eds.): AIDD 2024, LNCS 14894, pp. 132–148, 2025.
https://doi.org/10.1007/978-3-031-72381-0_11

of humans in the models [2]. The effects are particularly important in high-stakes decision-making processes that rely on machine learning as they allow users to judge results based on the predicted uncertainty quantification before basing critical decisions on the results [11]. Drug discovery is a complex field of research where experiments are time-consuming, expensive, and high-risk, therefore wrong decisions regarding which experiments to make can be highly wasteful [29]. Additionally, the early stages of drug discovery rely on modeling the complex chemical space where data availability is typically limited, another effect of the time-consuming and costly experiments needed to generate data. As such, there is a continuously increasing need to develop application-specific uncertainty quantification methods in molecular property prediction and modeling of quantitative structure-activity relationships (QSAR) [15].

Approaches that quantify uncertainty in machine learning for regression tasks can be classified into Bayesian learning [7], ensemble-based [12,25,36,38], distance-based [4,40], mean-variance-estimation [6,8,31], evidential learning [1], etc. Several recent efforts have been made to compare and benchmark the available methods on publicly available datasets related to molecular property prediction or QSAR modeling [10,16,18,23,42,47]. However, no consensus has been reached regarding a single method that consistently outperforms the other methods across evaluation metrics and tasks [48]. Hirschfeld et al. [18] stress the need for a more realistic evaluation, such as a temporal data split, to gain insights into the real implications and nuances between the approaches. Additionally, Yin et al. [47] point out that public benchmarks do not allow proper temporal evaluation as they lack relevant information and sufficient replications for reliable statistics.

Prior work that uses temporal evaluation on public data for molecular property prediction can be misleading [27]. The reason is that the available information regarding the time of data points in public data does not relate to the real evolution of experiments in a pharmaceutical company, which is what makes a temporal evaluation truly useful in real drug discovery. Earlier work on internal pharmaceutical assay-based data from Merck compares a temporal splitting strategy with random and structure-based splitting strategies [39]. Sheridan [39] concludes that the temporal option best approximates the true predictive performance, but they do not explore uncertainty quantification.

Uncertainty quantification can be disentangled to detail the underlying sources behind the uncertainty, which gives a more comprehensive understanding of the factors that contribute to the total predictive uncertainty. In machine learning, the two main sources of uncertainty can be derived from are the aleatoric and the epistemic parts [2,20,22]. Aleatoric uncertainty is the inherent stochastic variability in experiments, also considered irreducible as it cannot be reduced with additional data or changes to the model. Epistemic uncertainty includes all remaining sources, such as lack of knowledge and model limitations. The epistemic uncertainty can be reduced with additional data or changes to the model, but understanding which adjustments are needed requires further dissection of the predicted uncertainty [13]. Awareness of the aleatoric uncertainty in molecular property prediction can lead to better risk management by recognizing and quantifying the unpredictable nature of certain properties or parts of the chemical space [46]. Quantified epistemic uncertainty, on the other hand, can be used during drug discovery to guide the search through the chemical space by redirecting data collection [16]. If the parts of the epistemic uncertainty that relate to missing data or distribution shift can be effectively separated from the remaining model uncertainty, it can also aid in developing the machine learning model.

In this work, we provide a sought-after comparison of available methods for uncertainty quantification in a temporal evaluation of assay-based QSAR modeling for real pharmaceutical data. We focus the analysis on ensemble-based approaches that quantify predictive uncertainty and attempt to further disentangle the uncertainty between distributional uncertainty and model uncertainty such that the results are most useful in guiding the real-world search for new drugs. Additionally, we explore the effects of including lower-quality data through censored labels during training.

## 2    Methods

Our analysis has been performed on data from ten internal biological assays differentiated by the categories proposed by Heyndrickx et al. [17], namely Panel, Other, and ADME assays. The Panel category includes cross-project assays related to undesired off-target effects. The Other category includes on-target activity from project-specific assays. The ten assays presented in this work belong to these two categories. Larger assays of the ADME type, related to Absorption, Distribution, Metabolism, and Excretion, are left for future work. The respective distributions of observed experimental labels for each assay are shown in the bottom half of Fig. 1.

All but one of the assays model pIC50 values, while the Other 3 assay models pEC50. Due to the infeasibility of performing an unlimited number of experiments to find exact experimental results, such as pIC50 values, significant proportions of the data are provided as censored labels. Censored labels define a threshold below or above which the true results lie, e.g. the censored label $< 3$ pIC50 means that the true pIC50 value is below three. In some cases, the censored labels have been included in model training, as explained further in the following section. However, note that the available censored labels are highly imbalanced, as for all but two assays less than 1% of the censored labels are lower bound, i.e. $>$. The Other 2 assay has just above 1% of $>$ censored labels and the Other 4 assay has 2%, while the $<$ labels typically make up between 30–60% of each assay's total number of results. There are two assays without any censored labels, namely Other 5 and 7. Data points that are not censored are called observed labels in the remainder of this work.

Duplicated measurements for molecular compounds in the data are aggregated using the median of the result and the standard deviation is stored for later reference. Each molecular compound is then encoded with RDKit [26]

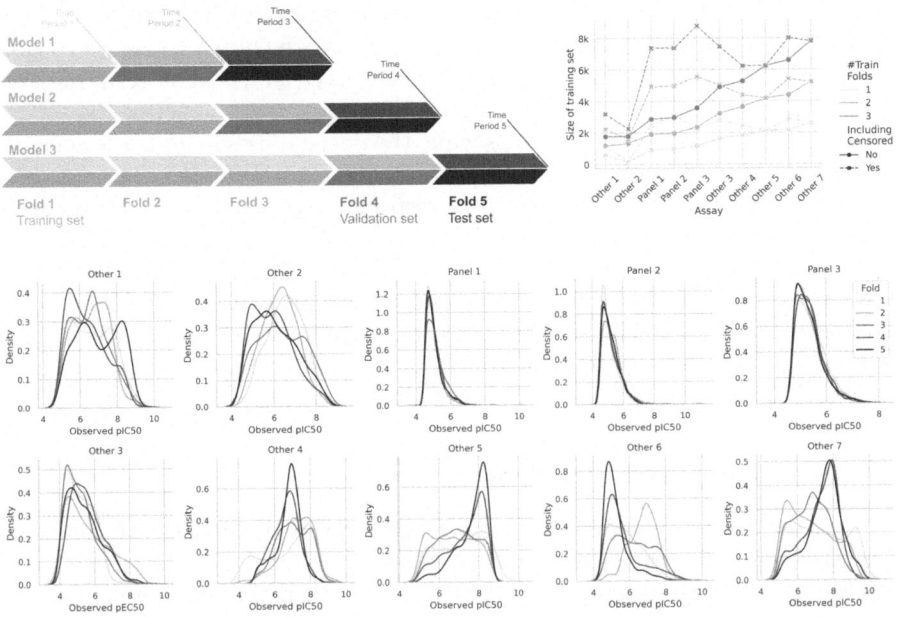

**Fig. 1. Five-fold temporal split.** (Upper left) Five folds and how they are used to create three temporal settings, each with more training data. For each setting, the first subsequent fold is used for validation and calibration, and the second subsequent fold is used for testing. (Upper right) Training data size for each assay and temporal setting, with and without including available censored labels. (Lower) Distribution of observed labels across the temporal folds for two example assays, one from each category.

from SMILES strings [43] to Morgan Fingerprints [30] of size 1024 and radius 2. Other. More advanced ways to encode molecular compounds exist, such as the graph-based ChemProp model [46] and the pre-trained language-based CDDD model [44]. Models based on the resulting embeddings from these neural network encoders have been compared and shown improvements in prior work [10,18,27]. Specifically, Dutschmanm et al. [10] showed that fingerprints perform best in combination with RF and are close second to CDDD in combination with neural networks. While the fingerprint representations are used in our study for simplicity, we encourage considering state-of-the-art, learned representations before deploying the proposed methods in practical applications.

**Temporal Split.** The main contribution of our work relates to evaluating the uncertainty quantification of molecular property prediction in a temporal evolving setting. As such, we simulate realistic assay-based modeling of pharmaceutical projects by splitting the data of each assay into five folds based on the date of the experiment. Where duplicated measurements were aggregated, the first experiment date of all measurements was used. The upper left panel in Fig. 1

illustrates the folds and resulting three settings that can be used to evaluate trained models as time evolves. The time intervals are chosen to create roughly equally sized folds regarding the number of observed labels. The resulting sizes of training sets for each assay are shown in the top right panel of Fig. 1. The solid lines show only observed results while the dashed lines include the censored labels. Note that the size of the setting with one train fold also corresponds to the size of the validation and test sets respectively, as individual, subsequent folds are used for these.

As previously mentioned, the lower part of Fig. 1 illustrates the distribution of observed labels in each fold of every assay. Note particularly, the shift in distributions between folds in the Other assays compared to the highly similar label distributions over time in the Panel assays. The assays are ordered according to the overall dataset size throughout this work.

## 2.1 Ensemble-Based Modeling

We compare three ensemble-based approaches for regression QSAR modeling of several internal biological assays. As such, we consider each assay $t$ as an individual single-task dataset $\mathcal{D}_t := \{(\mathbf{x}^n, y_t^n)\}_{n=1}^N$ of molecular compounds represented by a one-dimensional numerical embeddings $\mathbf{x}^n \in \mathbb{R}^e$ and continuous activity labels $y_t^n \in \mathbb{R}$. An ensemble is defined as a set of $K$ base estimators $\hat{y}_t^n = f(\mathbf{x}^n)$. We consider two base estimators, a decision tree regressor and a multi-layer perceptron (MLP), i.e. fully connected deep neural network. We take the average of the individual base estimators' predictions as the final prediction by the ensemble $\mu_t$ and define the variance of the predictions as an estimate of the predictive uncertainty $\sigma_t^2$, as follows

$$\mu_t(\mathbf{x}^n) = \frac{1}{K} \sum_{k=1}^K f_k(\mathbf{x}^n), \quad \sigma_t^2(\mathbf{x}^n) = \frac{1}{K} \sum_{k=1}^K (f_k(\mathbf{x}^n))^2 - (\mu_t(\mathbf{x}^n))^2. \quad (1)$$

The ensemble of decision tree regressors results in a Random Forest (RF) model [38] while we use the MLP base estimator to create a Deep Ensemble (DE) as proposed by Lakshminarayanan et al. [25] and an MC-Dropout model as proposed by Gal & Ghahramani [12]. Prior work has compared similar methods for variability in QSAR modeling [41]. The DE combines base 50 MLPs trained from different weight initialization whereas the MC model generates 500 samples from a single trained base MLP with dropout turned on during inference.

In a Bayesian framework, the uncertainty in model parameters $\omega$ results in the predictive uncertainty of the model $p(y_t^n|\mathbf{x}^n, \omega)$. The true posterior distribution of model parameters for a given dataset can be described as $p(\omega|\mathcal{D}_t)$, such that the predictive uncertainty of the Bayesian model average is defined by $p(y_t^n|\mathbf{x}^n, \mathcal{D}_t) = \int_\Omega p(y_t^n|\mathbf{x}^n, \tilde{\omega}) p(\tilde{\omega}|\mathcal{D}_t) d\tilde{\omega}$ [11,20]. As shown by both Lakshminarayanan et al. [25] and Gal & Ghahramani [11], the variance in ensemble predictions provides an approximation of the epistemic part of this true posterior distribution. Figure 2 gives an overview of the three ensemble-based methods considered in our work. The remainder of this section gives details about the training procedures used in the evaluation of the three methods.

**Training Details.** The Random Forest is implemented using scikit-learn [34] and the two neural network-based models are trained with PyTorch [32]. All models are initially trained with a Mean Squared Error (MSE) loss only on data points with observed labels. However, in addition, we include versions of the neural network-based models for which censored labels are also included in the training data. We denote these models as DE+ and MC+ in the result. Note that these extended models are not provided for the Other 5 and 7 assays, which do not include censored labels. Training these extended models requires adjustments to the loss function, as censored labels only give a one-sided view of the true result. We adopt the CensoredMSE defined by Arany et al. [3] with a one-sided squared error applied for the censored labels as follows

$$\mathcal{L}(\mathbf{x}^n, y_t^n) = \frac{1}{N} \sum_{n=1}^{N} \begin{cases} \min\left(y_t^n - \mu_t(\mathbf{x}^n), 0\right)^2, & \text{if censored label} < y_t^n, \\ (y_t^n - \mu_t(\mathbf{x}^n))^2, & \text{if observed label } y_t^n, \\ \max\left(y_t^n - \mu_t(\mathbf{x}^n), 0\right)^2, & \text{if censored label} > y_t^n. \end{cases} \quad (2)$$

To compare the models trained on censored labels fairly against the ones trained only on observed labels we only include the censored labels in the training sets. Thus, the validation and test sets are identical between the models. We believe this could hinder the censored models somewhat, especially due to the imbalance between lower and upper-bound labels.

We optimize the hyperparameters for each base estimator detailed in Table 1 of the Appendix for each assay and each temporal setting individually using a grid search based on the validation MSE loss. It would not be computationally feasible to optimize the DE model in terms of any score that incorporates the calibration of uncertainty estimates due to the large number of models that would need to be trained. Therefore, we do not consider this option for any of the models to ensure a fair comparison. However, such optimization schemes should be considered for practical applications.

**Evaluation.** While the MSE loss is used to evaluate the performance of the predictions made by the models, other metrics are required to evaluate the accuracy and calibration of the predicted uncertainties. We consider two types of ways to evaluate predicted uncertainty, ones that evaluate only the accuracy or calibration of the uncertainty and ones that evaluate predictive performance intertwined with how well-calibrated the predicted uncertainty is. A detailed way to evaluate the predicted uncertainties by themselves is by comparing the confidence-based calibration curve to the identity function which corresponds to perfect calibration [16,19,42,45]. The confidence-based calibration curve is obtained by computing the z% confidence interval (CI) for every predicted uncertainty in the test set. Next, the observed fraction of errors within each CI is calculated for several expected fractions between 0 and 1.

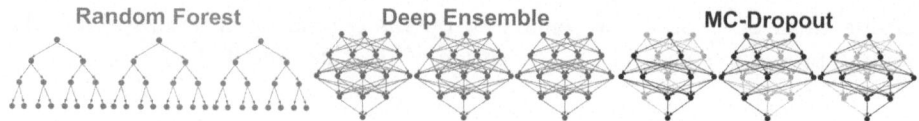

**Fig. 2. Ensemble-based models.** Three approaches to ensemble-based modeling including uncertainty quantification.

Furthermore, the Gaussian Negative Log Likelihood (NLL) [49] and the Expected Normalized Calibration Error (ENCE) [28] are two global metrics that evaluate the intertwined predictive performance and calibration of uncertainties. The Gaussian NLL is defined as,

$$
\text{NLL} = \frac{1}{2N} \sum_{n=1}^{N} \left( \ln(2\pi) + \ln(\sigma_t^2(\mathbf{x}^n)) + \frac{(y_t^n - \mu_t(\mathbf{x}^n))^2}{\sigma_t^2(\mathbf{x}^n)} \right). \tag{3}
$$

The ENCE metric is derived from the error-based calibration plot proposed by Levi et al. [28] which is made from a binned representation of the Root MSE and the Root Mean Variance (RMV), i.e. predicted uncertainty. Computationally, the errors and corresponding predicted uncertainties are ordered based on increasing predicted uncertainty and split into a set $\mathcal{B}$ of bins. For each bin $b$ of size $|b|$ the RMSE and RMV are calculated as,

$$
\text{RMSE}_b = \sqrt{\frac{1}{|b|} \sum_{i \in b} (y_t^i - \mu_t(\mathbf{x}^i))^2}, \quad \text{RMV}_b = \sqrt{\frac{1}{|b|} \sum_{i \in b} \sigma_t^2(\mathbf{x}^i)}. \tag{4}
$$

Finally, the bins are summarized to give the ENCE metric as follows,

$$
\text{ENCE} = \frac{1}{|\mathcal{B}|} \sum_{b \in \mathcal{B}} \frac{|\text{RMSE}_b - \text{RMV}_b|}{\text{RMV}_b}. \tag{5}
$$

Several additional metrics have been proposed and used to evaluate uncertainty estimates in drug discovery applications, such as Spearman's Rank Correlation Coefficient between predicted uncertainties and corresponding errors [10,18,42,47]. However, this score has been criticized due to the stochasticity and unreliability of the result [35]. Statistically, a data point with high predicted uncertainty can still result in a prediction with low error and vice versa. Therefore, we discard the metric from our analysis.

**Recalibration.** Several post hoc alternatives have been proposed to recalibrate predicted uncertainties by ensemble-based models [21,28,35], as the original estimates have been found to underestimate the epistemic uncertainty [9,37]. Janet et al. [21] recalibrate the uncertainty estimates based on a maximum-likelihood estimation strategy on the NLL, while Levi et al. [28] propose a re-scaling of the predicted uncertainty based on the NLL similar to temperature scaling [14].

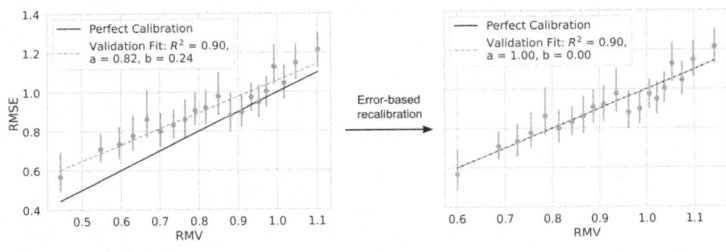

**Fig. 3. Error-based recalibration.** Linear recalibration of uncertainty estimates based on the validation set.

Most recently, Rasmussen et al. [35] instead proposed to recalibrate the predicted uncertainty using the fit of the RMSE versus RMV curve described above as the error-based calibration plot. The latter is the strategy that we adopt in this work and Fig. 3 illustrates an example of a recalibration on the validation set of one of our datasets. A linear regression is fitted to the binned RMSE versus RMV results on the validation set, resulting in parameters $a_{\mathrm{val}}$ for the slope and $b_{\mathrm{val}}$ for the intercept. During inference the predicted standard deviation is then shifted according to $\sigma_{\mathrm{cal}} = a_{\mathrm{val}} \cdot \sigma + b_{\mathrm{val}}$.

## 3   Experiments

In the experimental setup, we first analyze and compare the performance of the models averaged over ten repeated experiments on all assays and temporal settings. The global evaluation scores are shown in Fig. 4 and the confidence-based calibration curves are shown in Fig. 5. We then provide a more in-depth case study of the predictions by one of the best-performing models on the Other 6 assay, which exhibits a particularly challenging distribution shift in terms of both the feature and label space. Here, we illustrate how the predicted uncertainties relate to the distribution shift in the feature space and suggest how the model's predictions could have practical implications for future decisions in the given drug discovery project.

**Model Comparison.** Figure 4 presents an overview of the MSE and recalibrated NLL and ENCE scores. Note that the recalibration step only affects the predicted uncertainties and therefore does not affect the MSE. In the figure, the models can be compared in several ways: 1) as the training set size increases over time for each assay with increasing #Train folds, 2) as the overall size of the assay increases, going from smallest assays in the left-most columns to larger assays in the right-most columns, 3) in terms of the varying amounts of label shifts between the Panel and Other assays, or 4) in terms of the different metrics.

The first observable trend is that predictive performance is higher for the Panel assays compared to the Other assays. This is not surprising given the

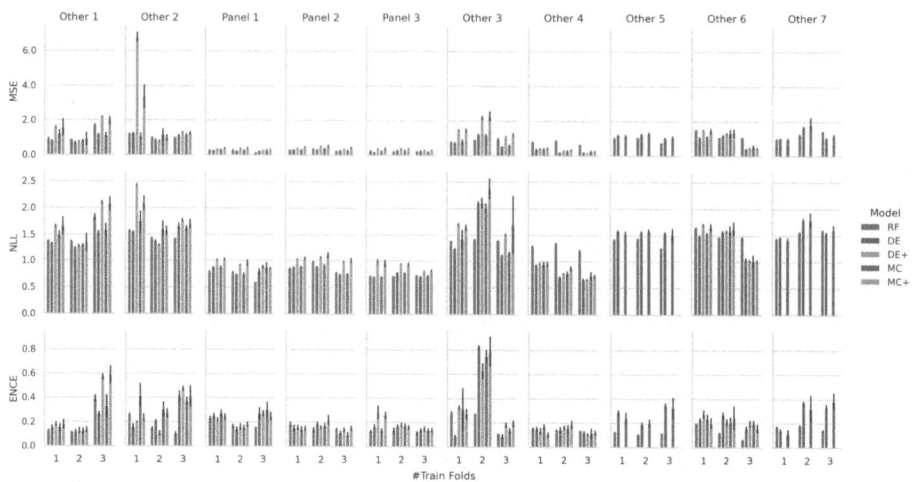

**Fig. 4. Benchmarking overview.** Results for each assay and temporal setting averaged over ten repeated experiments. DE+ and MC+ are trained with censored labels as supplementary lower-quality data. However, these models do not apply to Other 5 and 7 as they do not include censored labels.

constant distribution over time as illustrated in Fig. 1. A similar trend can be observed in the NLL but not in terms of ENCE. As the Gaussian NLL includes the squared error term, a likely conclusion is that distribution shifts do not generally hurt the calibration of uncertainty estimates. This conclusion is also reasonable as the predictive uncertainty from ensemble-based approaches model specifically the epistemic uncertainty which should cover distribution shifts. In general, the ranking of the methods from the MSE scores are often the same in the NLL while they can vary in terms of the ENCE. For example, for the Other 1 assay the DE is always among the best models for all three temporal settings in terms of the MSE and NLL scores, while in terms of the ENCE score, it is outperformed by the RF model in the first two temporal settings.

For the most part, the performances of the two MLP-based models are usually indistinguishable from each other for the cases trained with and without censored data respectively. On the contrary, there are no general trends regarding whether the RF model or the MLP-based models are best. This changes depending on the assay, metric, and even temporal setting. However, the versions of the neural network-based models trained with supplementary censored labels, DE+ and MC+, do not generally improve the predictive performance or the calibration of uncertainty estimates over their respective base versions. Only one instance occurs where the DE+ is better than the DE model and all other models for all three scores, namely the Other 2 model trained on 2 folds. However, as this result is not consistent across the other two temporal settings of the assay, it is more likely the result of statistical variability. The non-competitive results with

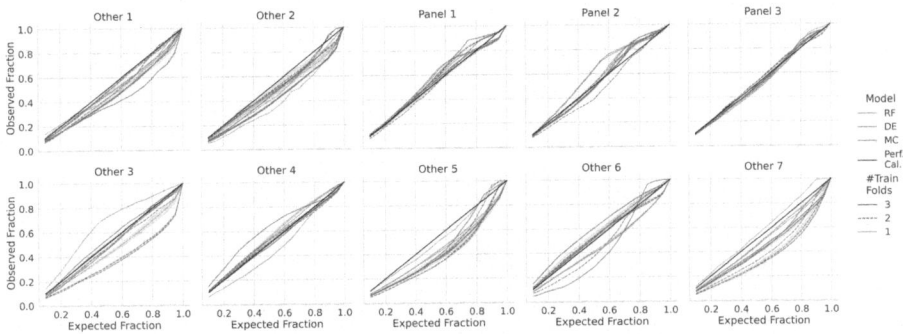

**Fig. 5. Confidence-based calibration over time.** Discrete visualization of the observed fraction of results for each expected confidence interval based on the predicted uncertainty. The black, solid lines illustrate perfect calibration.

the censored models require further analysis, but we believe it could be due to the uneven nature of the censored labels available. As described in Sect. 2, the vast majority of the censored labels are upper bound ($<$). Given that all models are evaluated only on observed labels for a fair comparison, the imbalance in the censored labels may shift the models' understanding of the label distributions.

In light of the overall poor performance of the models trained with censored data, we have omitted these models from the confidence-based calibration curves presented in Fig. 5. The curves are shown for each assay and temporal setting with error bands illustrating the confidence from the ten repeated experiments. A majority of calibration curves are not far away from being perfectly calibrated. This indicates that most models produce useful uncertainty estimates. The possibility of such intuitive interpretations of the calibration curves is not as easily derived from the scores presented in Fig. 4. The reason for this is that the ENCE score is unbounded, such that it can be hard to determine whether achieved scores are useful or not. For the calibration curves in Fig. 4 it is clear that they are significantly closer to being perfectly calibrated than to the extreme cases of completely over- or under-calibration. On the other hand, it is harder to compare the models and temporal settings in terms of the calibration curves, as many of the curves are indistinguishable. However, in practical applications where perhaps a particular confidence is of interest, a closer evaluation of the calibration curves can be crucial to distinguish between the models.

**Case Study.** Finally, we provide a practical case study of one of the Other assays, Other 6, which exhibits a particularly challenging evolution of the data throughout time. Our case study aims to test the top-performing model from the model comparison above in this demanding setting to determine in detail how well the predictive uncertainties perform, and how individual predictions can be used in practice to impact future decisions of the drug discovery project.

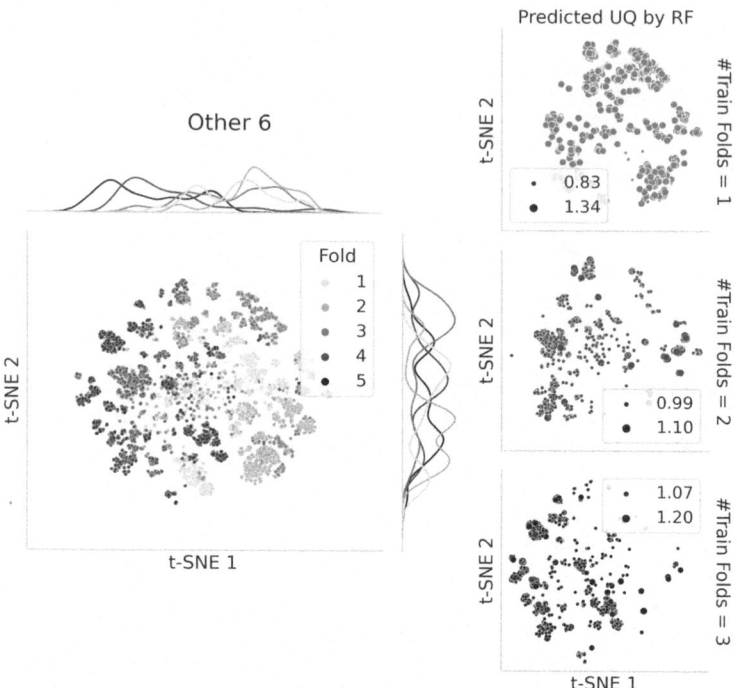

**Fig. 6. Practical temporal evaluation.** A t-SNE projection of the Other 6 assay, colored by temporal fold. The left panel illustrates the full dataset, where a distribution shift can be seen throughout time. The remaining panels in the right column, show individual test sets with predicted uncertainty by the RF model presented as the size of data points.

The leftmost part of Fig. 6 illustrates the feature space of the compounds tested on the assay decomposed to two t-SNE projections and colored by the five temporal folds. A clear distribution shift in the feature space can be observed in the t-SNE projection where the second fold tends more toward the bottom right corner of the feature space and the last two folds shift drastically to the left side of the plot. Similarly, highly varying label distributions were seen between the same folds in the lower part of Fig. 1 in Sect. 2. Also, the label distribution does not shift continuously over time, but instead first shifts greatly toward higher pIC50 values in the second fold and then back toward more extreme lower values by the last two folds. In the remaining three plots to the right in Fig. 6, the t-SNE projections of each test set, i.e. folds 3, 4, and 5, are repeated separately. Here, the size of the data points is determined by the recalibrated predicted uncertainty of the RF model trained on the three temporal settings respectively, i.e. with an increasing number of training folds. The RF model is chosen for this analysis due to being best-performing on the Other 6 assay in terms of the

ENCE score. Note that the legends of these plots detail the respective minimum and maximum predicted uncertainties on the given test set.

We observe that the model trained on the least amount of data namely on only the first fold and tested on fold 3, seen in the top panel of the right column in Fig. 6, indicates overall high uncertainty for most test compounds. A likely explanation is that the amount of training data was insufficient for the model to learn from, meaning that it overfitted and could not generalize well to the test compounds. The described scenario is also corroborated by the relatively poor MSE score seen for RF trained on one fold of assay Other 6 in Fig. 4 compared to the same model trained on two and three folds respectively. For the models trained on two and three folds, the span of predicted uncertainties is notably much smaller, 0.11 and 0.13, compared to the first model, 0.51. As a result, we can observe more distinct patterns in the predicted uncertainty between different regions of the feature space. The regions with high uncertainty predicted by the model trained on two folds seem uncorrelated with proximity to training data. However, when the third and final training fold is included, it is clear that the clusters with the highest predicted uncertainty are also located furthest away from the training data. The same trend is reflected in the ENCE scores presented in Fig. 4 where the calibration error of the model trained on two folds is significantly worse than the one achieved by the model trained on three folds.

Given the distribution shift present in the feature space, and that the ensemble-based model's predicted uncertainty accounts for epistemic uncertainty, it follows our expectation that the distribution shift should be reflected in the estimated uncertainty. As such, our analysis provides empirical evidence to support this claim, but it also illustrates that the uncertainty estimates cover additional sources of uncertainty related to the model itself such as overfitting. It is important to understand all sources of uncertainty when basing future high-stakes decisions on them, such as in drug discovery. Considering the identified cases in this case study, we provide practical suggestions on how the identified sources can impact the continued drug discovery process. If overfitting is determined, such as through overall high uncertainty estimates and low performance seen in the model train on one fold, the modeling requires overall more data before deployment. Another alternative would be to reconsider the choice of model, but our temporal split shows that the RF continues to be the best choice in the future when more data is included. When distribution shifts are instead identified, such as seen later in the given project for the model trained on three folds, more data exploration is needed in the chemical spaces where the uncertainty estimates are high before deployment.

Further research is necessary to disentangle the sources of epistemic uncertainty between distribution shifts and other model-related sources, such that more reliable measurements of these situations can be obtained. One alternative approach would be to quantify the distribution shift using other means, either with distance-based approaches, such as the average Tanimoto similarity [40] between an inference compound and compounds in the training set, or the interpretable method proposed by Kulinski and Inouye [24]. Additionally, more advanced pre-training procedures can be used, that are trained to incorporate distribution shift more effectively [5]. After the distribution shift has been independently quantified, the predicted epistemic uncertainty could be re-evaluated such that the remaining model uncertainty is disentangled from this information.

## 4   Conclusions

In this comparison between three ensemble-based uncertainty quantification approaches evaluated temporally on data from multiple biological assays, we have shown varying results between the assays emphasizing the impact of individual assay characteristics on predictive outcomes. No single model was consistently best across evaluation metrics or assays, but some conclusions could be drawn for particular assays. Specifically, we analyzed the results in light of the varying presence of shifts in label distributions and feature space distributions in the assays over time. While doing so we found that predictive performance and calibration of uncertainty can be robust and reliable for assays without distribution shifts and that the method can be used to identify data points outside of the training distribution when distribution shifts are present. As such, we give insights and provide practical advice on how uncertainty estimates by ensemble-based models can be used to impact future decision-making in high-stakes situations such as drug discovery. Incorporating lower-quality data in the form of censored labels did not yield improvements in the predictive performance of the models. Suggestions were given as to why this could be the case, such as the uneven nature of the censored labels and the evaluation strategy. Future work can explore other ways to include the censored labels or extend the analysis to other modeling approaches that allow censored labels, such as Censored Quantile Regression [33]. Overall, this study has gained valuable insights into how distribution shift affects uncertainty quantification in assay-based QSAR modeling, which can impact real-world pharmaceutical drug discovery.

**Acknowledgements.** We thank our colleagues and reviewers for their valuable feedback, especially Susanne Winiwarter at AstraZeneca in Gothenburg for her advice and guidance during the data preparation. This study was partially funded by the European Union's Horizon 2020 research and innovation programme under the Marie Skłodowska-Curie Actions grant agreement "Advanced machine learning for Innovative Drug Discovery (AIDD)" No. 956832.

**Disclosure of Interests.** The authors have no competing interests to declare that are relevant to the content of this article.

## Appendix

Table 1 presents the hyperparameters explored in the model selection for the RF model and the base MLP used for both the DE and MC models. A grid search was used to find the optimal hyperparameters for every assay and temporal setting based on the validation MSE loss. Additionally, the MLPs were trained using the Adam optimizer with a weight decay of 0.0005, the learning rate was reduced when plateauing with a patience of 50 epochs, and a batch size of 64 was used.

**Table 1. Model selection.** Considered hyperparameter space for model selection of RF and base MLP during grid search based on validation MSE loss.

| Base Model | Hyperparameter | Explored space |
|---|---|---|
| RF | n_estimators | {50, 100, 250, 500, 1000} |
| | min_samples_leaf | {2, 10, 0.25, 0.5, 0.75} |
| | min_samples_split | {1, 25, 50, 100, 250, 500} |
| MLP | Learning rate | {0.00005, 0.0001, 0.0005, 0.001} |
| | Scheduler Factor | {0.1, 0.5} |
| | Number of hidden layers | {2, 3, 4} |
| | Hidden dimension | {64, 128, 256, 512} |
| | Decreasing dimension | {False, True} |
| | Dropout | {0, 0.25, 0.5, 0.75} |

# References

1. Amini, A., Schwarting, W., Soleimany, A., Rus, D.: Deep evidential regression. In: Advances in Neural Information Processing Systems, vol. 33, pp. 14927–14937. Curran Associates, Inc. (2020)
2. Apostolakis, G.: The concept of probability in safety assessments of technological systems. Science **250**(4986), 1359–1364 (1990)
3. Arany, A., Simm, J., Oldenhof, M., Moreau, Y.: SparseChem: fast and accurate machine learning model for small molecules. arXiv preprint arXiv:2203.04676 (2022)
4. Berenger, F., Yamanishi, Y.: A distance-based boolean applicability domain for classification of high throughput screening data. J. Chem. Inf. Model. **59**(1), 463–476 (2018)
5. Bertolini, M., Clevert, D.A., Montanari, F.: Explaining, evaluating and enhancing neural networks' learned representations. In: International Conference on Artificial Neural Networks, pp. 269–287. Springer, Cham (2023). https://doi.org/10.1007/978-3-031-44192-9_22
6. Bishop, C.M.: Mixture Density Networks. Technical report. Aston University, Birmingham (1994)
7. Blundell, C., Cornebise, J., Kavukcuoglu, K., Wierstra, D.: Weight uncertainty in neural network. In: International Conference on Machine Learning, pp. 1613–1622. PMLR (2015)

8. Choi, S., Lee, K., Lim, S., Oh, S.: Uncertainty-aware learning from demonstration using mixture density networks with sampling-free variance modeling. In: 2018 IEEE International Conference on Robotics and Automation (ICRA), pp. 6915–6922. IEEE (2018)
9. D'Angelo, F., Fortuin, V.: Repulsive deep ensembles are bayesian. In: Advances in Neural Information Processing Systems, vol. 34, pp. 3451–3465. Curran Associates, Inc. (2021)
10. Dutschmann, T.M., Kinzel, L., Ter Laak, A., Baumann, K.: Large-scale evaluation of k-fold cross-validation ensembles for uncertainty estimation. J. Cheminf. **15**(1), 49 (2023)
11. Gal, Y.: Uncertainty in Deep Learning. Ph.D. thesis, Department of Engineering, University of Cambridge (2016)
12. Gal, Y., Ghahramani, Z.: Dropout as a bayesian approximation: representing model uncertainty in deep learning. In: International Conference on Machine Learning, pp. 1050–1059. PMLR (2016)
13. Gruber, C., Schenk, P.O., Schierholz, M., Kreuter, F., Kauermann, G.: Sources of Uncertainty in Machine Learning–A Statisticians' View. arXiv preprint arXiv:2305.16703 (2023)
14. Guo, C., Pleiss, G., Sun, Y., Weinberger, K.Q.: On calibration of modern neural networks. In: International Conference on Machine Learning, pp. 1321–1330. PMLR (2017)
15. Hansch, C., Fujita, T.: p-$\sigma$-$\pi$ Analysis. A Method for the Correlation of Biological Activity and Chemical Structure. J. Am. Chem. Soc. **86**(8), 1616–1626 (1964)
16. Heid, E., McGill, C.J., Vermeire, F.H., Green, W.H.: Characterizing uncertainty in machine learning for chemistry. J. Chem. Inf. Model. **63**(13), 4012–4029 (2023)
17. Heyndrickx, W., et al.: MELLODDY: Cross-pharma Federated Learning at Unprecedented Scale Unlocks Benefits in QSAR without Compromising Proprietary Information. J. Chem. Inf, Model (2023)
18. Hirschfeld, L., Swanson, K., Yang, K., Barzilay, R., Coley, C.W.: Uncertainty quantification using neural networks for molecular property prediction. J. Chem. Inf. Model. **60**(8), 3770–3780 (2020)
19. Hubschneider, C., Hutmacher, R., Zöllner, J.M.: Calibrating uncertainty models for steering angle estimation. In: 2019 IEEE Intelligent Transportation Systems Conference (ITSC), pp. 1511–1518. IEEE (2019)
20. Hüllermeier, E., Waegeman, W.: Aleatoric and epistemic uncertainty in machine learning: an introduction to concepts and methods. Mach. Learn. **110**, 457–506 (2021)
21. Janet, J.P., Duan, C., Yang, T., Nandy, A., Kulik, H.J.: A quantitative uncertainty metric controls error in neural network-driven chemical discovery. Chem. Sci. **10**(34), 7913–7922 (2019)
22. Kendall, A., Gal, Y.: What uncertainties do we need in bayesian deep learning for computer vision? In: Advances in Neural Information Processing Systems, vol. 30. Curran Associates, Inc. (2017)
23. Kim, Q., Ko, J.H., Kim, S., Park, N., Jhe, W.: Bayesian neural network with pretrained protein embedding enhances prediction accuracy of drug-protein interaction. Bioinformatics **37**(20), 3428–3435 (2021)
24. Kulinski, S., Inouye, D.I.: Towards explaining distribution shifts. In: International Conference on Machine Learning, pp. 17931–17952. PMLR (2023)
25. Lakshminarayanan, B., Pritzel, A., Blundell, C.: Simple and scalable predictive uncertainty estimation using deep ensembles. In: Advances in Neural Information Processing Systems, vol. 30. Curran Associates, Inc. (2017)

26. Landrum, G.: RDKit: Open-Source Cheminformatics (2006). https://doi.org/10.5281/zenodo.6961488, http://www.rdkit.org
27. Lenselink, E.B., et al.: Beyond the hype: deep neural networks outperform established methods using a ChEMBL bioactivity benchmark set. J. Cheminf. **9**(1), 1–14 (2017)
28. Levi, D., Gispan, L., Giladi, N., Fetaya, E.: Evaluating and calibrating uncertainty prediction in regression tasks. Sensors **22**(15), 5540 (2022)
29. Mervin, L.H., Johansson, S., Semenova, E., Giblin, K.A., Engkvist, O.: Uncertainty quantification in drug design. Drug Discovery Today **26**(2), 474–489 (2021)
30. Morgan, H.L.: The generation of a unique machine description for chemical structures - a technique developed at chemical abstracts service. J. Chem. Doc. **5**(2), 107–113 (1965)
31. Nix, D.A., Weigend, A.S.: Estimating the mean and variance of the target probability distribution. In: Proceedings of 1994 IEEE International Conference on Neural Networks (ICNN'94), vol. 1, pp. 55–60. IEEE (1994)
32. Paszke, A., et al.: PyTorch: an imperative style, high-performance deep learning library. In: Advances in Neural Information Processing Systems, vol. 32. Curran Associates, Inc. (2019)
33. Pearce, T., Jeong, J.H., Jia, Y., Zhu, J.: Censored quantile regression neural networks for distribution-free survival analysis. In: Advances in Neural Information Processing Systems, vol. 35, pp. 7450–7461. Curran Associates, Inc. (2022)
34. Pedregosa, F., et al.: Scikit-learn: machine learning in python. J. Mach. Learn. Res. **12**, 2825–2830 (2011)
35. Rasmussen, M.H., Duan, C., Kulik, H.J., Jensen, J.H.: Uncertain of uncertainties? a comparison of uncertainty quantification metrics for chemical data sets. J. Cheminf. **15**(1), 121 (2023)
36. Scalia, G., Grambow, C.A., Pernici, B., Li, Y.P., Green, W.H.: Evaluating scalable uncertainty estimation methods for deep learning-based molecular property prediction. J. Chem. Inf. Model. **60**(6), 2697–2717 (2020)
37. Schweighofer, K., Aichberger, L., Ielanskyi, M., Klambauer, G., Hochreiter, S.: Quantification of Uncertainty with Adversarial Models. In: Advances in Neural Information Processing Systems, vol. 36. Curran Associates, Inc. (2023)
38. Sheridan, R.P.: Three useful dimensions for domain applicability in QSAR models using random forest. J. Chem. Inf. Model. **52**(3), 814–823 (2012)
39. Sheridan, R.P.: Time-split cross-validation as a method for estimating the goodness of prospective prediction. J. Chem. Inf. Model. **53**(4), 783–790 (2013)
40. Sheridan, R.P., Feuston, B.P., Maiorov, V.N., Kearsley, S.K.: Similarity to molecules in the training set is a good discriminator for prediction accuracy in QSAR. J. Chem. Inf. Comput. Sci. **44**(6), 1912–1928 (2004)
41. Tetko, I.V., et al.: Critical Assessment of QSAR Models of Environmental Toxicity Against Tetrahymena Pyriformis: Focusing on Applicability Domain and Overfitting by Variable Selection. J. Chem. Inf. Model. **48**(9), 1733–1746 (2008)
42. Wang, D., et al.: A hybrid framework for improving uncertainty quantification in deep learning-based QSAR regression modeling. J. Cheminf. **13**(1), 1–17 (2021)
43. Weininger, D.: SMILES, a Chemical Language and Information System. 1. Introduction to Methodology and Encoding Rules. J. Chem. Inf. Comput. Sci. **28**(1), 31–36 (1988)
44. Winter, R., Montanari, F., Noé, F., Clevert, D.A.: Learning continuous and data-driven molecular descriptors by translating equivalent chemical representations. Chem. Sci. **10**(6), 1692–1701 (2019)

45. Yang, C.I., Li, Y.P.: Explainable uncertainty quantifications for deep learning-based molecular property prediction. J. Cheminf. **15**(1), 13 (2023)
46. Yang, K., et al.: Analyzing learned molecular representations for property prediction. J. Chem. Inf. Model. **59**(8), 3370–3388 (2019)
47. Yin, T., Panapitiya, G., Coda, E.D., Saldanha, E.G.: Evaluating uncertainty-based active learning for accelerating the generalization of molecular property prediction. J. Cheminf. **15**(1), 105 (2023)
48. Yu, J., Wang, D., Zheng, M.: Uncertainty quantification: can we trust artificial intelligence in drug discovery? iScience **25**(8), 104814 (2022)
49. Zadrozny, B., Elkan, C.: Obtaining calibrated probability estimates from decision trees and naive bayesian classifiers. In: International Conference on Machine Learning, pp. 609–616. PMLR (2001)

**Open Access** This chapter is licensed under the terms of the Creative Commons Attribution 4.0 International License (http://creativecommons.org/licenses/by/4.0/), which permits use, sharing, adaptation, distribution and reproduction in any medium or format, as long as you give appropriate credit to the original author(s) and the source, provide a link to the Creative Commons license and indicate if changes were made.

The images or other third party material in this chapter are included in the chapter's Creative Commons license, unless indicated otherwise in a credit line to the material. If material is not included in the chapter's Creative Commons license and your intended use is not permitted by statutory regulation or exceeds the permitted use, you will need to obtain permission directly from the copyright holder.

# Deep Bayesian Experimental Design
# for Drug Discovery

Muhammad Arslan Masood[1,2](✉) [iD], Tianyu Cui[2] [iD], and Samuel Kaski[2,3] [iD]

[1] Drug Discovery Data Sciences, Janssen Pharmaceutica NV, Turnhoutseweg 30,
2340 Beerse, Belgium
[2] Department of Computer Science, Aalto University, Espoo, Finland
arslan.masood@aalto.fi
[3] Department of Computer Science, University of Manchester, Manchester, UK
https://www.aalto.fi/en/department-of-computer-science

**Abstract.** In drug discovery, prioritizing compounds for testing is an important task. Active learning can assist in this endeavor by prioritizing molecules for label acquisition based on their estimated potential to enhance in-silico models. However, in specialized cases like toxicity modeling, limited dataset sizes can hinder effective training of modern neural networks for representation learning and to perform active learning. In this study, we leverage a transformer-based BERT model pretrained on millions of SMILES to perform active learning. Additionally, we explore different acquisition functions to assess their compatibility with pretrained BERT model. Our results demonstrate that pretrained models enhance active learning outcomes. Furthermore, we observe that active learning selects a higher proportion of positive compounds compared to random acquisition functions, an important advantage, especially in dealing with imbalanced toxicity datasets. Through a comparative analysis, we find that both BALD and EPIG acquisition functions outperform random acquisition, with EPIG exhibiting slightly superior performance over BALD. In summary, our study highlights the effectiveness of active learning in conjunction with pretrained models to tackle the problem of data scarcity.

**Keywords:** Drug Discovery · Active learning · Bayesian · BERT

## 1 Introduction and Background

Drug design is a complex process, with costs exceeding $4 billion and a decade of development time required to bring a new drug to market (Schlander et al., 2021). Despite this investment, a vast majority of drugs never make it to clinical

© The Author(s) 2025
D.-A. Clevert et al. (Eds.): AIDD 2024, LNCS 14894, pp. 149–159, 2025.
https://doi.org/10.1007/978-3-031-72381-0_12

trials and of those drugs that do enter clinical trials a staggering 90% of drugs fail (Sun et al., 2022), with 50% of failures attributed to unexpected human toxicity (Van Norman, 2019). Traditional toxicological studies rely on animal models at the preclinical stage, yet these models face limitations in reliability, time, and ethical concerns, with their translational relevance to humans remaining uncertain (Raies and Bajic, 2016).

The adoption of the 3R principles (Replace, Reduce, Refine) to curtail animal testing has catalyzed the development of in vitro methods for toxicological assessment of new compounds (Choudhuri et al., 2018). In the early phases of drug discovery, multiple cytotoxicity assays measure the impact of chemical compounds on cellular structure and function, providing early indications of potential tissue and organ toxicity (Ballantyne, 2006; Tabernilla et al., 2021).

Well-designed in vitro experiments can reduce the reliance on animal testing. An *experiment* is a systematic procedure aimed at collecting scientific data to test hypotheses or generate new ones. Common experimental designs include completely randomized experiments or randomized block testing (Festing, 2001).

In contexts such as high throughput screening (HTS) and toxicity assays, where exhaustive search is infeasible due to the vast number of possible combinations, efficient experimental design is paramount (Niedz and Evens, 2016). It's simply not feasible to test every drug against each target. Bayesian experimental design (BeD) emerges as a powerful tool in this regard, reducing the required number of experiments (Khan et al., 2023). BeD achieves this by providing hypothetical experimental options based on the outcomes of previous ones, thereby potentially curtailing costs and expediting the drug discovery process (Daly et al., 2019; Bader et al., 2023).

In-silico methods are often used in conjunction with in-vitro studies to model the behaviour of biological systems by leveraging available experimental data (Merino-Casallo et al., 2018; Abd El Hafez et al., 2022). Bayesian methods have been applied to select the optimal parameters of the in-vitro experiments (Pauwels et al., 2014; Johnston et al., 2016), parameters estimation of mechanistic models (Merino-Casallo et al., 2018; Demetriades et al., 2022), estimating drug synergies (Cremaschi et al., 2019; Rønneberg et al., 2021), and computing in-vitro dose response curves (Hennessey et al., 2010). It still remains unclear which experiment to conduct next in order to obtain the most informative data point for inclusion in the subsequent iteration of training, aimed at enhancing the overall performance of these in-silico models. Addressing this challenge, we borrowed Bayesian methods for experimental design from the computer vision community and applied to model toxicity endpoints.

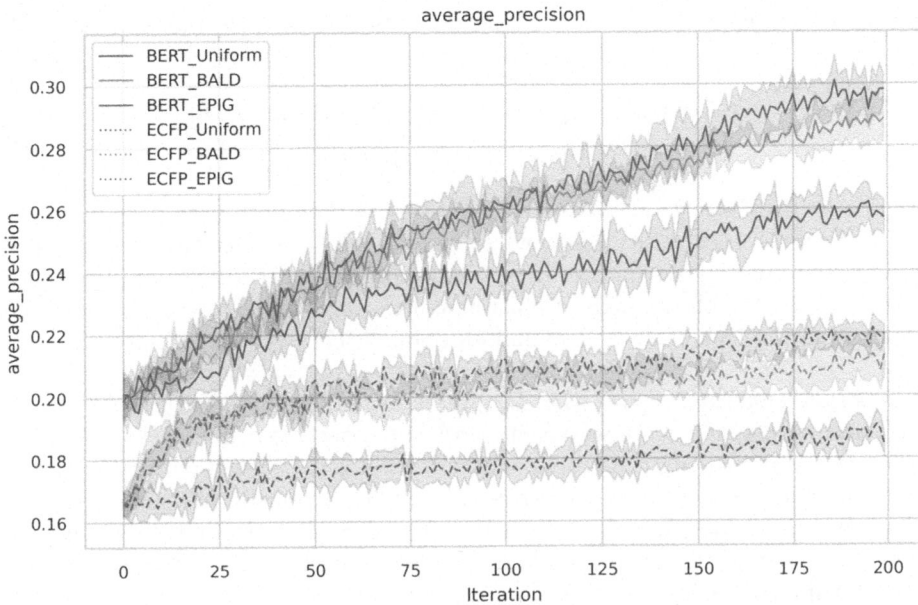

**Fig. 1.** Mean average precision across 12 tasks of Tox21 dataset. Active learning with pretrained BERT features outperforms models trained on ECFP. Furthermore, BALD and EPIG acquisition functions select more informative samples than uniform (random) sampling, with EPIG demonstrating a slight superiority over BALD

## 2    Methods

### 2.1    Bayesian Active Learning

We first consider fully supervised learning tasks, e.g., estimating molecular properties, using a probabilistic model with likelihood function $p(y|\boldsymbol{x}, \phi)$, where $\boldsymbol{x}$ is an input, $y$ is the output, and $\phi$ is the parameter of the model $f(\boldsymbol{x}; \phi)$ which has a prior distribution $p(\phi)$ and a posterior $p(\phi|\mathcal{D})$ given a labelled training set $\mathcal{D} = \{(\boldsymbol{x}_i, y_i)\}_{i=1}^N$. In active learning or experimental design (Rainforth et al., 2024), we have access to another unlabelled set $\mathcal{D}_u = \{(\boldsymbol{x}_i^u)\}_{i=1}^{N_u}$ and select which labels to acquire when training the model $f(\boldsymbol{x}; \phi)$ by maximizing an acquisition function that captures the expected utility of acquiring the label $y_s$ of the selected input $\boldsymbol{x}_s$. Then the new labelled data $(\boldsymbol{x}_s^u, y_s)$ is incorporated into the training set $\mathcal{D} = \mathcal{D} \bigcup \{(\boldsymbol{x}_s^u, y_s)\}$ and the probabilistic model, i.e., the posterior $p(\phi|\mathcal{D})$, is updated accordingly.

*Acquisition Function: BALD* One popular acquisition function is Bayesian Active Learning by Disagreement (BALD) (Houlsby et al., 2011), which is the expected information gain, measured by the reduction in Shannon entropy, of the model parameter $\phi$ from labelling $\boldsymbol{x}$ across all possible realisations of its label $y$ given by $p(y|\boldsymbol{x}, \mathcal{D})$. Specifically, we have

$$\begin{aligned}
\text{BALD}(\boldsymbol{x}) &= \mathbb{E}_{y \sim p(y|\boldsymbol{x}, \mathcal{D})} \left[ \text{H}[\phi|\mathcal{D}] - \text{H}[\phi|\boldsymbol{x}, y, \mathcal{D}] \right] \\
&= \text{H}[y|\boldsymbol{x}, \mathcal{D}] - \mathbb{E}_{\phi \sim p(\phi|\mathcal{D})} \left[ \text{H}[y|\boldsymbol{x}, \phi] \right]
\end{aligned} \tag{1}$$

with the optimal design $\boldsymbol{x}^\star = \arg\max_{\boldsymbol{x}} \text{BALD}(\boldsymbol{x})$. The first term in BALD measures the total uncertainty on $\boldsymbol{x}$ while the second term measures its aleatoric uncertainty, i.e., the irreducible uncertainty from observational noise. Therefore, BALD selects $\boldsymbol{x}$ with the highest epistemic uncertainty, i.e., the reducible uncertainty from the lack of data (Kendall and Gal, 2017).

*Acquisition Function: EPIG* BALD targets a global uncertainty reduction on the parameter space $\phi$. However, in most supervised learning tasks, users are interested in improving the model accuracy on a target set $p(\boldsymbol{x}_*)$, e.g., the test set. Therefore, recent work (Smith et al., 2023a) claimed that an acquisition function, Expected Predictive Information Gain (EPIG), explicitly reducing the model output uncertainty on random samples from $p(\boldsymbol{x}_*)$ is more effective than BALD in improving the model performance. Specifically, as discussed and defined in (Smith et al., 2023b) EPIG$(\boldsymbol{x}) =$

$$\mathbb{E}_{p(\boldsymbol{x}_*)} \left[ \text{H}[y_*|\boldsymbol{x}_*, \mathcal{D}] - \mathbb{E}_{p(y|\boldsymbol{x}, \mathcal{D})} \left[ \text{H}[y_*|\boldsymbol{x}_*, y, \boldsymbol{x}] \right] \right] \tag{2}$$

is *expected* reduction of the "expected predictive uncertainty" over the *target input distribution* $p(\boldsymbol{x}_*)$ by observing the label of $\boldsymbol{x}$. Intuitively, compared with BALD which reduces the parameter uncertainty globally, EPIG only reduces the parameter uncertainty that reduces model output uncertainty on $p(\boldsymbol{x}_*)$.

*Semi-supervised Active Learning (SSAL).* In the fully supervised scenario, the model $f(\boldsymbol{x}; \phi)$ only learns from the labelled dataset $\mathcal{D}$. This is inefficient in active learning because the labelled dataset for training is limited initially, and active learning has to collect more data to learn a good input manifold, which is required to estimate the uncertainty of downstream tasks (Smith et al., 2024). This is particularly challenging in the chemical space, where the input manifold is nontrivial (Zhou et al., 2019). Therefore, researchers proposed semi-supervised active learning (SSAL) approaches (Zhang et al., 2019; Hao et al., 2020) to learn the representations of input molecules using both labelled and unlabeled data and conduct active learning on the representation space with the labelled data. However, the available molecules in most public molecular property datasets are still limited (ranging from hundreds to thousands), even without labels.

In this paper, we propose to use molecular representations from a pretrained self-supervised learning model. Specifically, we encoded the molecular SMILES sequences into corresponding embeddings, utilizing a large transformer model MolBERT, pretrained on 1.6 million SMILES via masking, alongside physicochemical properties (Fabian et al., 2020) . The embedding of each SMILES sequence is a pooled output from the pretrained MolBERT with dimension 764.

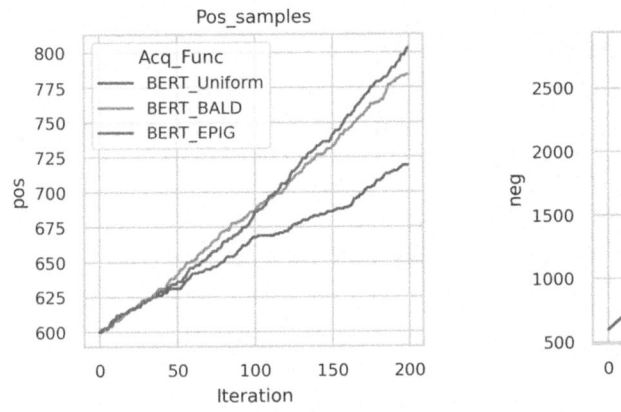
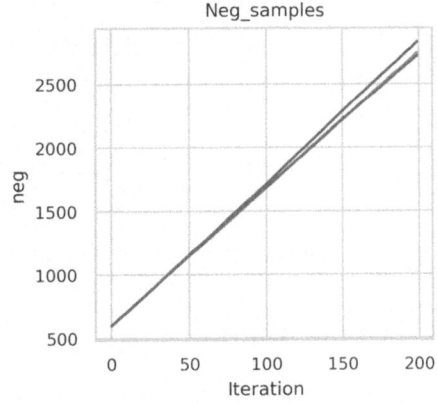

**Fig. 2.** Sum of samples across 12 tasks of Tox21 dataset. EPIG and BALD are acquiring more positive sample than random acquisition

We employed these embeddings from MolBERT to train a fully connected (i.e., MLP) head. This strategy allowed us to leverage a significant volume of molecule data, offering particular benefits for conducting active learning on relatively small datasets.

## 2.2  Practical Bayesian Neural Networks

In this work, we use a Bayesian neural network to account for the model uncertainty. According to recent research on dropout variational inference (Gal and Ghahramani, 2016), a practical Bayesian neural network for a wide variety of architectures can be obtained by simply training a neural network with dropout (MC dropout), and interpreting this as being equivalent to variational inference (Blei et al., 2017). The uncertainty is then estimated by using the multiple forward-passing with different dropout masks. Although the uncertainty from MC dropout is often underestimated, it has been a popular choice for Bayesian active learning with neural networks and shows promise on real-world datasets (Gal et al., 2017; Rakesh and Jain, 2021).

This neural network architecture consists of an input-hidden-output layers, where $x_0$ is initialized as the input features $x$, which can be either BERT features (in the semi-supervised AL) or binary fingerprints (in the supervised AL). We utilize dropout for uncertainty estimation, batch normalization for training stability, and the rectified linear unit (ReLU) activation function as the default activation. Additionally, the network incorporates a skip connection, merging the input and output of the hidden layer, enhancing information flow. Finally, the output layer generates logits, which can be transformed into probabilities by passing through a sigmoidal activation function.

$$x_0 = x \quad \text{BERT features or ECFP}$$
$$x_\ell = \text{Dropout}(\text{ReLU}(\text{BatchNorm}(W_\ell x_0 + \mathbf{b}_\ell)))$$
$$\tilde{x}_{\ell+1} = \text{BatchNorm}(W_{\ell+1} x_\ell + \mathbf{b}_{\ell+1}) \tag{3}$$
$$x_{\ell+1} = \text{Dropout}(\text{ReLU}(x_\ell + \tilde{x}_{\ell+1}))$$
$$x_{out} = W_{\ell+2} x_{\ell+1} + \mathbf{b}_{\ell+1}$$

The hyper-parameters of this model are given in Table 1.

## 3    Experiments

### 3.1    Dataset

*Tox21.* The Tox21 dataset, or Toxicology in the 21st Century dataset, is a publicly available dataset used in the field of computational toxicology (Richard et al., 2021). The Tox21 dataset consists of a large collection of chemical compounds, each of which is associated with various types of toxicity outcomes. These outcomes are typically measured using high-throughput screening assays to evaluate the potential toxic effects of the compounds. The dataset provides a quantitative assessment (in form of binary labels) of toxicity of $\approx$ 8000 compounds in 12 different toxicity pathways.

**Table 1.** Hyperparameters used of BNN and training

|          | Hyperparameter          | Values             |
|----------|-------------------------|--------------------|
| BNN      | Activation              | [ReLU]             |
|          | Batch normalization     | [True]             |
|          | Skip connection         | [True]             |
|          | Input layer             | [768, 1024]        |
|          | hidden layer dim        | [128]              |
|          | Number of hidden layers | [1]                |
|          | Dropout probability     | [0.3]              |
| Training | Optimizer               | [Adam]             |
|          | Learning rate           | $[10^{-3}]$        |
|          | Weight decay            | [1e-2]             |
|          | Scheduler               | [CosineAnnealingLR]|
|          | T-max (LR cycle)        | [10]               |
|          | Batch size              | [16]               |
|          | Epochs                  | [110]              |
|          | num. Forward pass       | [20]               |

The Tox21 dataset is widely used as a benchmark in the development of in silico toxicology models. In this dataset, 6.24% measurements are active (ranges from 2% to 12%), 73% are inactive, while 20.56% are missing values as shown in Fig. 3.

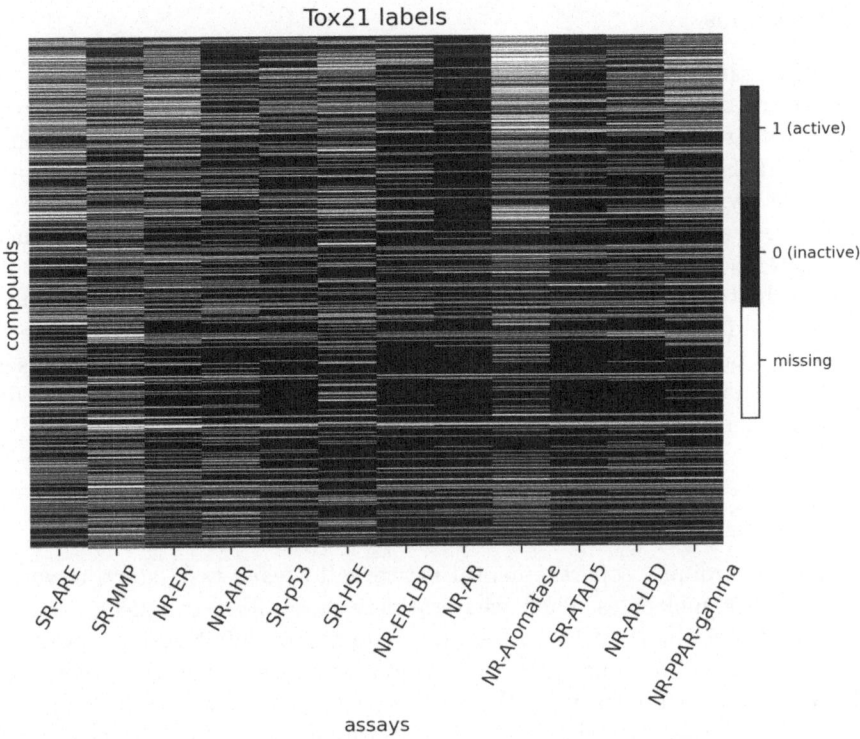

**Fig. 3.** The output space of Tox21, displaying active compounds in red, inactive compounds in blue, and missing data points in white. (Color figure online)

## 3.2   Data Splitting

*Test, Train Set.* For the better of evaluation of generalization, we employed scaffold splitting with 80:20 ratio to create distinct training and testing sets. Scaffold splitting partitions a molecular dataset according to core structural motifs identified by the Bemis-Murcko scaffold representation (Bemis and Murcko, 1996), prioritizing larger groups while ensuring that the train and test sets do not share identical scaffolds. The testset for all the experiments is identical.

*Initial and Pool Sets.* A balanced initial set was constructed by randomly selecting 100 molecules from the training set, with equal representation of positive and negative instances. Subsequently, a pool set was generated by excluding the initial set from the training set.

### 3.3 Baselines

We consider three acquisition functions, random, BALD, and EPIG (Sect. 2.1), and two learning paradigms, supervised active learning (AL) and semi-supervised active learning (SSAL). In SSAL, we use the BERT features pretrained on 1.6 million SMILES, and in AL, we use ECFP, or Extended-Connectivity Fingerprints, directly. ECFP is a method used in cheminformatics to represent molecular structures as binary fingerprints, capturing structural information by encoding the presence or absence of substructural features within a specified radius around each atom. Through iterative traversal of the molecular structure, unique substructural fragments are identified and hashed into a fixed-length bit vector, generating a binary fingerprint where each bit indicates the presence or absence of a specific substructural fragment. We encoded each molecule into a fixed 1024-dimensional binary vector using a radius of 6.

## 4 Results and Discussions

We began by training separate neural networks for each task, starting with an initial set of 100 molecules. Then, we iteratively chose the next molecule based on acquisition functions (BALD, EPIG, and random) for 200 iterations, evaluating the test set after each round. Our study compared active learning strategies using both ECFP and BERT features. We repeated this process with 5 different seeds, showing the average precision (AUPR) performance evolution across iterations (Fig. 1). Notably, active learning with pretrained BERT features outperformed models trained on ECFP. Additionally, BALD and EPIG acquisition functions consistently selected more informative samples than uniform (random) sampling, with EPIG showing a slight edge over BALD. Many learning algorithms face challenges in effectively learning from imbalanced datasets, where the dominance of the majority class can overwhelm the learning process. As illustrated in Fig. 2, our analysis demonstrates that both EPIG and BALD consistently acquire a higher proportion of positive samples compared to random acquisition. This observation holds particular significance in the modeling of toxicity datasets.

**Acknowledgments.** The authors acknowledge financial support from the European Union's Horizon 2020 research and innovation program under the Marie Skłodowska-Curie grant agreement No 956832, "Advanced Machine learning for Innovative Drug Discovery" (AIDD).

**Disclosure of Interests.** The authors have no competing interests to declare that are relevant to the content of this article.

# References

Abd El Hafez, M.S., et al.: Characterization, in-silico, and in-vitro study of a new steroid derivative from Ophiocoma dentata as a potential treatment for COVID-19. Sci. Rep. **12**(1), 5846 (2022). ISSN 2045-2322. https://doi.org/10.1038/s41598-022-09809-2, https://www.nature.com/articles/s41598-022-09809-2. Publisher: Nature Publishing Group

Bader, J., Narayanan, H., Arosio, P., Leroux, J.C.: Improving extracellular vesicles production through a Bayesian optimization-based experimental design. Eur. J. Pharm. Biopharm. **182**, 103–114 (2023). ISSN 0939-6411. https://doi.org/10.1016/j.ejpb.2022.12.004, https://www.sciencedirect.com/science/article/pii/S0939641122002983

BALLANTYNE, B.: Local and systemic ophthalmic pharmacology and toxicology of organophosphate and carbamate anticholinesterases. In: Toxicology of Organophosphate & Carbamate Compounds, pp. 423–445. Elsevier, 2006. ISBN 978-0-12-088523-7. https://doi.org/10.1016/B978-012088523-7/50032-6, https://linkinghub.elsevier.com/retrieve/pii/B9780120885237500326

Bemis, G.W., Murcko, M.A.: The properties of known drugs. 1. molecular frameworks. J. Med. Chem. **39**(15), 2887–2893 (1996). ISSN 0022-2623. https://doi.org/10.1021/jm9602928. Publisher: American Chemical Society

Blei, D.M., Kucukelbir, A., McAuliffe, J.D.: Variational inference: a review for statisticians. J. Am. Stat. Assoc. **112**(518), 859–877 (2017)

Choudhuri, S., Patton, G.W., Chanderbhan, R.F., Mattia, A., Klaassen, C.D.: From classical toxicology to Tox21: some critical conceptual and technological advances in the molecular understanding of the toxic response beginning from the last quarter of the 20th century. Toxicol. Sci. **161**(1), 5–22 (2018). ISSN 1096-6080, 1096-0929. https://doi.org/10.1093/toxsci/kfx186, https://academic.oup.com/toxsci/article/161/1/5/4102075

Cremaschi, A., Frigessi, A., Taskén, K., Zucknick, M.: A Bayesian approach to study synergistic interaction effects in in-vitro drug combination experiments. arXiv:1904.04901 (2019)

Daly, A.J., Stock, M., Baetens, J.M., De Baets, B.: Guiding mineralization co-culture discovery using bayesian optimization. Environ. Sci. Technol. **53**(24), 14459–14469 (2019). ISSN 0013-936X. https://doi.org/10.1021/acs.est.9b05942. Publisher: American Chemical Society

Demetriades, M., et al.: Interrogating and quantifying in vitro cancer drug pharmacodynamics via agent-based and bayesian monte carlo modelling. Pharmaceutics **14**(4), 749 (2022). ISSN 1999-4923. https://doi.org/10.3390/pharmaceutics14040749, https://www.mdpi.com/1999-4923/14/4/749. Number: 4 Publisher: Multidisciplinary Digital Publishing Institute

Fabian, B., Edlich, T., Gaspar, H., Segler, M., Meyers, J., Fiscato, M.: Molecular representation learning with language models and domain-relevant auxiliary tasks. arXiv:2011.13230 (2020)

Festing, M.F.: Guidelines for the design and statistical analysis of experiments in papers submitted to ATLA. Alternatives to Laboratory Animals (2001). https://doi.org/10.1177/026119290102900409, https://journals.sagepub.com/doi/10.1177/026119290102900409. Publisher: SAGE PublicationsSage UK: London, England

Gal, Y., Ghahramani, Z.: Dropout as a Bayesian approximation: representing model uncertainty in deep learning. In: Proceedings of the 34th International Conference on Machine Learning, pp. 1050–1059 (2016)

Gal, Y., Islam, R. and Ghahramani, Z.: Deep Bayesian active learning with image data. In: International Conference on Machine Learning, pp. 1183–1192. PMLR (2017)

Hao, Z., et al.: ASGN: an active semi-supervised graph neural network for molecular property prediction. In: Proceedings of the 26th ACM SIGKDD International Conference on Knowledge Discovery & Data Mining, pp. 731–752 (2020)

Hennessey, V.G., Rosner, G.L., Bast Jr, R.C., Chen, M.Y.: A Bayesian approach to dose-response assessment and synergy and its application to in vitro dose-response studies. Biometrics, **66**(4), 1275–1283 (2010). ISSN 0006-341X. https://doi.org/10.1111/j.1541-0420.2010.01403.x

Houlsby, N., Huszár, F., Ghahramani, Z., Lengyel, M.: Bayesian active learning for classification and preference learning. arXiv preprint arXiv:1112.5745 (2011)

Johnston, S.T., Ross, J.V., Binder, B.J., McElwain, D.S., Haridas, P., Simpson, M.J.: Quantifying the effect of experimental design choices for in vitro scratch assays. J. Theor. Biol. **400**, 19–31 (2016). ISSN 0022-5193. https://doi.org/10.1016/j.jtbi.2016.04.012, https://www.sciencedirect.com/science/article/pii/S0022519316300406

Kendall, A., Gal, Y.: What uncertainties do we need in Bayesian deep learning for computer vision? In: Advances in Neural Information Processing Systems, vol. 30 (2017)

Khan, A., et al.: Toward real-world automated antibody design with combinatorial Bayesian optimization. Cell Rep. Methods **3**(1) (2023.) ISSN 2667-2375. https://doi.org/10.1016/j.crmeth.2022.100374, https://www.cell.com/cell-reports-methods/abstract/S2667-2375(22)00276-4. Publisher: Elsevier

Merino-Casallo, F., Gomez-Benito, M.J., Juste-Lanas, Y., Martinez-Cantin, R., Garcia-Aznar, J.M.: Integration of in vitro and in silico models using bayesian optimization with an application to stochastic modeling of mesenchymal 3D cell migration. Front. Phys. **9**, 1246 (2018). ISSN 1664-042X. https://doi.org/10.3389/fphys.2018.01246, https://www.ncbi.nlm.nih.gov/pmc/articles/PMC6142046/

Niedz, R.P., Evens, T.J.: Design of Experiments (DOE)—history, concepts, and relevance to in vitro culture. Vitro Cell. Dev. Biol. Plant **52**(6), 547–562 (2016). https://doi.org/10.1007/s11627-016-9786-1

Pauwels, E., Lajaunie, C., Vert, J.P.: A Bayesian active learning strategy for sequential experimental design in systems biology. BMC Syst. Biol. **8**(1), 102 (2014). ISSN 1752-0509. https://doi.org/10.1186/s12918-014-0102-6, https://bmcsystbiol.biomedcentral.com/articles/10.1186/s12918-014-0102-6

Raies, A.B., Bajic, V.B.: In silico toxicology: computational methods for the prediction of chemical toxicity: computational methods for the prediction of chemical toxicity. Wiley Interdiscip. Rev. Comput. Mol. Sci. **6**(2), 147–172 (2016). ISSN 17590876. https://doi.org/10.1002/wcms.1240, https://onlinelibrary.wiley.com/doi/10.1002/wcms.1240

Rainforth, T., Foster, A., Ivanova, D.R., Bickford Smith, F.: Modern Bayesian experimental design. Stat. Sci. **39**(1), 100–114 (2024)

Rakesh, V., Jain, S.: Efficacy of Bayesian neural networks in active learning. In: Proceedings of the IEEE/CVF Conference on Computer Vision and Pattern Recognition, pp. 2601–2609 (2021)

Richard, A.M., et al.: The Tox21 10K compound library: collaborative chemistry advancing toxicology. Chem. Res. Toxicol. 34(2), 189–216 (2021). ISSN 0893-228X, 1520-5010. https://doi.org/10.1021/acs.chemrestox.0c00264, https://pubs.acs.org/doi/10.1021/acs.chemrestox.0c00264

Rønneberg, L., Cremaschi, A., Hanes, R., Enserink, J.M., Zucknick, M.: Bayesynergy: flexible Bayesian modelling of synergistic interaction effects in in vitro drug combination experiments. Briefings Bioinform. **22**(6), bbab251 (2021). ISSN 1477-4054. https://doi.org/10.1093/bib/bbab251

Schlander, M., Hernandez-Villafuerte, K., Cheng, C.-Y., Mestre-Ferrandiz, J., Baumann, M.: How much does it cost to research and develop a new drug? a systematic review and assessment. Pharmacoeconomics **39**(11), 1243–1269 (2021). https://doi.org/10.1007/s40273-021-01065-y

Smith, F.B., Kirsch, A., Farquhar, S., Gal, Y., Foster, A., Rainforth, T.: Prediction-oriented Bayesian active learning. In: International Conference on Artificial Intelligence and Statistics, pp. 7331–7348. PMLR (2023)

Smith, F.B., Kirsch, A., Farquhar, S., Gal, Y., Foster, A., Rainforth, T.: Prediction-oriented Bayesian active learning. arXiv:2304.08151v1 (2023)

Smith, F.B., Foster, A., Rainforth, T.: Making better use of unlabelled data in Bayesian active learning. In: International Conference on Artificial Intelligence and Statistics, pp. 847–855. PMLR (2024)

Sun, D., Gao, W., Hu, H. and Zhou, S.: Why 90% of clinical drug development fails and how to improve it? Acta Pharmaceutica Sinica. B **12**(7), 3049–3062 (2022). ISSN 2211-3835. https://doi.org/10.1016/j.apsb.2022.02.002, https://www.ncbi.nlm.nih.gov/pmc/articles/PMC9293739/

Tabernilla, A., et al.: In vitro liver toxicity testing of chemicals: a pragmatic approach. Int. J. Mol. Sci. **22**(9), 5038 (2021). ISSN 1422-0067. https://doi.org/10.3390/ijms22095038, https://www.mdpi.com/1422-0067/22/9/5038

Van Norman, G.A.: Phase II trials in drug development and adaptive trial design. JACC: Basic Transl. Sci. **4**(3), 428–437 (2019). ISSN 2452302X. https://doi.org/10.1016/j.jacbts.2019.02.005, https://linkinghub.elsevier.com/retrieve/pii/S2452302X19300658

Zhang, Y., et al.: Bayesian semi-supervised learning for uncertainty-calibrated prediction of molecular properties and active learning. Chem. Sci. **10**(35), 8154–8163 (2019)

Zhou, Z., Kearnes, S., Li, L., Zare, R.N., Riley, P.: Optimization of molecules via deep reinforcement learning. Sci. Rep. **9**(1), 10752 (2019)

**Open Access** This chapter is licensed under the terms of the Creative Commons Attribution 4.0 International License (http://creativecommons.org/licenses/by/4.0/), which permits use, sharing, adaptation, distribution and reproduction in any medium or format, as long as you give appropriate credit to the original author(s) and the source, provide a link to the Creative Commons license and indicate if changes were made.

The images or other third party material in this chapter are included in the chapter's Creative Commons license, unless indicated otherwise in a credit line to the material. If material is not included in the chapter's Creative Commons license and your intended use is not permitted by statutory regulation or exceeds the permitted use, you will need to obtain permission directly from the copyright holder.

# Abstracts from the AIDD Workshop

# Cartography-Driven Molecular Generation with ChemSpace Atlas

Fanny Bonachera$^{(\boxtimes)}$ , Mikhail Volkov , Dragos Horvath , Gilles Marcou ,
Olga Klimchuk , and Alexandre Varnek

Laboratoire de Chémoinformatique, UMR7140 CNRS/UniStra, University of
Strasbourg, Strasbourg, France
f.bonachera@unistra.fr
https://infochim.u-strasbg.fr

**Abstract.** The ChemSpace Atlas is a user-friendly tool developed to
answer the challenges of drug discovery in the context of Big Data,
with the use of Generative Topographic Mapping. Here, we present a
new addition to the functionalities offered by the ChemSpace Atlas: an
autoencoder-based *de novo* molecular generator.

**Keywords:** Chemoinformatics · Big Data · Deep Learning

## 1 ChemSpace Atlas Navigator

Nowadays, the evolution of combinatorial chemistry has led to a significant
increase in compound library size, making the process of searching for good drug
candidates more challenging. The *Structure generation module* of the ChemSpace
Atlas Navigator tool[1] was developed in answer of these novel challenges of "Big
Data"-driven drug discovery.

The tool is meant to respond to key requests by medicinal chemists: not only
must it be easy to navigate through chemical space and visualize compound similarities, but the researcher must also be able to predict and visualize properties
of said compounds, as well as their biological activities and polypharmacologi-
cal profiles. It supports analogue searching and structural analysis, all of these
features within the scope of chemical Big Data.

In the ChemSpace Atlas navigator [5], Generative Topographic Mapping
(GTM) represents the chemical space as fuzzy-logical 2-dimensional maps. GTM
landscapes can be colored in different ways, to represent comparisons of libraries,
or quantitative structure-activity/property (QSA/PR) analyses. Several universal GTMs (uGTM), based on different ISIDA chemical information-rich descriptors, are implemented in the ChemSpace Atlas navigator, providing different

---

[1] https://chematlas.chimie.unistra.fr/chemblactivity/chematlas_userspace/.

© The Author(s) 2025
D.-A. Clevert et al. (Eds.): AIDD 2024, LNCS 14894, pp. 163–166, 2025.
https://doi.org/10.1007/978-3-031-72381-0

perspectives on the different chemical spaces. Moreover, the hierarchical zooming approach (hGTM) makes it possible to "zoom" on different areas of the maps, up to 4 levels of zoom, and access a detailed view of the content of each libraries.

Currently, the ChemSpace Atlas contains thousands of GTMs, organized in a hierarchical structure. The different atlases proposed are based on ChEMBL activity profiling, natural products (from the COCONUT and ZINC databases), different types of compounds compiled from ChEMBL and COCONUT and commercial databases (ZINC).

The current development aims however to go beyond existing compound libraries, opening ChemSpace Atlas to the realm of possible structures, by coupling a *de novo* molecular generator to the mapping tool. These maps rely on expert-designed ISIDA fragment counts, which - unlike autoencoder latent vectors - cannot be directly decoded into structure, an attention-based conditional variational autoencoder (ACoVAE) [1] was implemented. This approach uses its distinct latent space representation, but relies on molecular descriptor vector "seeds" as "conditions" to bias the decoding process towards structures likely to correspond to the given seeds. The architecture is an innovation based on the model developed by Lin et al. [2]. The ACoVAE model has been trained to generate SMILES from chemical descriptors, namely from the in-house ISIDA descriptors [4].

The model trains with a list of SMILES and their associated ISIDA descriptors. The training follows 3 steps:

- First, a random latent vector distribution is generated from the SMILES in the training dataset, using a Gated Recurrent Unit (GRU)-based encoder. At this stage, we aim for an hyperspherical distribution with a mean of 0 and a variance of 1.
- Second, the initial descriptor vectors are transformed into conditional latent vectors, using a Grouped Linear Transformation layer (GLT).
- Finally, the conditional and random latent vectors are converted into SMILES strings using an Autoregressive Multihead Attention (AMA) decoder.

Then, during inference, a specific vector corresponding to the desired structure is used, along with random latent vectors sampled from the hyperspherical distribution previously generated. The supplied descriptor vector acts as the "condition" while the sampled vectors define the latent space. The model generates desired SMILES based on these 2 inputs, and, thanks to the random factor, produces alternative SMILES at each inference even if the same specific vector is used.

Using this ACoVAE model, two new functionalities have been added to the atlas of ChEMBL activities, one important chapter in ChemSpace Atlas:

- "Molecular" seed descriptor vector can be taken directly from user-input, relevant compounds, with the ACoVAE model generating new compounds that "mimic" the former. Running this in the ChemSpace context has however the benefit of allowing both seed compounds and analogues in the rich context provided by all the molecules in the databases behind: property landscapes predicting the proficiency to hit certain targets, be bioavailable, soluble, original with respect to commercial alternatives, etc. A pop-up window displays the list of valid structures and allows the projection of these new compounds directly on the chosen landscape, as illustrated in Fig. 1.

- "Node" seed descriptor vectors are obtained from the GTM nodes within an area of user interest on the landscape. For example, the coloration of a landscape makes it possible to easily identify areas containing numerous actives against a specific ChEMBL target. Therefore, when the user clicks on a node from one of these areas, descriptor vectors are sampled using a Gaussian distribution around the selected node, to generate a list of "seed" vectors for the ACoVAE. A list of compounds is then designed by the ACoVAE and displayed in a pop-up window. As for the previous functionality, it is possible in one click, to project the list of newly generated compounds into the same landscape of interest, and evaluate their predicted activity based on the area in which they are projected.

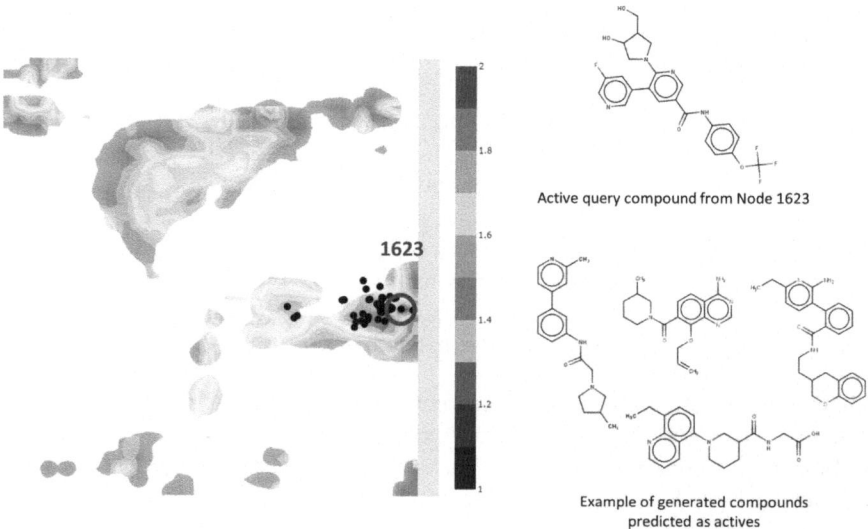

**Fig. 1.** On the left: Projection of the compounds generated from a query compound originating from node 1623 (black dots) on the activity landscape for the target CHEMBL1862. (Red: actives, blue: inactives). On the right: The query compound used as seed (top) and example of compounds generated using ACoVAE and projected on the landscape in the active zone around node 1623 (bottom). (Color figure online)

By default, four instances of the ACoVAE are launched in parallel on the ChemSpace Atlas server, to generate in total 40 new compounds. This process creates a variety of SMILES outputs for each combination of descriptor and random latent vectors. The validity of these generated SMILES is assessed using our internal standardization software based on the Indigo API [3], which filters out any erroneous sequences, and only valid SMILES are displayed to the user. This approach is effective in generating compounds that not only predict to have desired activities, but are also novel, whilst some display promising levels of drug-likeness and synthetic accessibility.

# References

1. Bort, W.E.A.: Inverse QSAR: reversing descriptor-driven prediction pipeline using attention-based conditional variational autoencoder. J. Chem. Inf. Model. **62**(22), 5471–5484 (2022)
2. Lin, Z.E.A.: Variational transformers for diverse response generation (2020, preprint). https://doi.org/10.48550/arXiv.2003.12738
3. Pavlov, D.E.A.: Indigo: universal cheminformatics API. J. Cheminformatics **3**(P4) (2011)
4. Ruggiu, F.E.A.: ISIDA property-labelled fragment descriptors. Mol. Inf. **29**(12), 855–868 (2010)
5. Zabolotna, Y.E.A.: Chemspace atlas: multiscale chemography of ultralarge libraries for drug discovery. J. Chem. Inf. Model. **62**(18), 4537–4548 (2022)

# Improving Route Development Using Convergent Retrosynthesis Planning

Paula Torren-Peraire[1,2](✉) [ID], Jonas Verhoeven[1] [ID], Dorota Herman[1] [ID], Hugo Ceulemans[1] [ID], Igor V. Tetko[2] [ID], and Jörg K. Wegner[3] [ID]

[1] In-Silico Discovery, Janssen Research and Development, Janssen Pharmaceutica N.V, Beerse, Belgium
ptorrenp@its.jnj.com
[2] Institute of Structural Biology, Molecular Targets and Therapeutics Center, Helmholtz Zentrum München, Neuherberg, Germany
[3] In-Silico Discovery, Janssen Research and Development, Janssen Research and Development LLC, Cambridge, USA

**Abstract.** Retrosynthesis consists of recursively breaking down a target molecule to produce a synthesis route composed of easily accessible building blocks. In recent years, computer-aided synthesis planning methods have allowed a greater exploration of potential synthesis routes, combining state-of-the-art machine-learning methods with chemical knowledge. However, these methods are generally developed to produce linear routes from a singular product to a set of proposed building blocks and are not designed to leverage potential shared paths between targets. Routes proposed from such developed methods do not encompass the real-world use case in medicinal chemistry where one often seeks to synthesize sets of target compounds in a library mode, ideally converging into a shared retrosynthetic path with respect to advanced intermediate compounds.

Using a graph-based processing pipeline, we explore Johnson & Johnson Electronic Laboratory Notebooks (J&J ELN) and publicly available datasets to identify complex routes with multiple products sharing common intermediates, producing convergent synthesis routes. We find that over 70% of all reactions are involved in convergent synthesis, covering over 80% of all projects in the case of J&J ELN data.

We introduce a novel planning approach to develop convergent synthesis routes, which can search multiple products and intermediates simultaneously as an extension of state-of-the-art machine learning single-step retrosynthesis models, enhancing the overall efficiency and practical applicability of retrosynthetic planning. We evaluate the multi-step synthesis planning approach using the extracted convergent routes and observe that solvability is generally very high across those routes, being able to identify a convergent route for over 90% of the test routes and showing an individual compound solvability of over 98%. Moreover, we find that using the convergent search approach allows us to further improve the accuracy of the suggested retrosynthetic routes, as compared to proposing individual routes.

© The Author(s) 2025
D.-A. Clevert et al. (Eds.): AIDD 2024, LNCS 14894, pp. 167–168, 2025.
https://doi.org/10.1007/978-3-031-72381-0

**Acknowledgments.** This study was partially funded by the European Union's Horizon 2020 research and innovation programme under the Marie Skłodowska-Curie Actions grant agreement "Advanced machine learning for Innovative Drug Discovery (AIDD)" No. 956832.

# Abstract Submission: Cross Multimodal Learning of Cell Painting and Transcriptomics Data

Son V. Ha[1,2(✉)], Steffen Jaensch[1], Maciej Kandula[1], Dorota Herman[1], Paul Czodrowski[2], and Hugo Ceulemans[1]

[1] Janssen Pharmaceutica N.V., a Johnson and Johnson company, Beerse, Belgium
sjaensch@its.jnj.com
[2] Department of Physical Chemistry, Johannes Gutenberg University, Mainz, Germany

## 1 Background

Multimodal learning is a subfield of machine learning that aims to process and relate information from multiple modalities [3]. This field has recently seen increased interest in drug discovery [4].

Cell Painting data (CP) is a type of morphological profiling, where a compound is characterized by the morphological changes in cells. Transcriptomics data (TX) is a type of gene expression profiling, where a compound is characterized by the expression change in genes. These two profiling methods can be viewed as two modalities that describe the same cell state. Hence, Multimodal learning techniques can potentially be used to capture the correspondences between them, potentially enhance understanding of biology.

In this work, we explore Cross Modality Feature Learning for CP and TX data. It is a setting where data from both modalities is available during feature learning, but only one modality is provided during supervised training and testing for downstream tasks. We find this setting particularly suitable for multimodal learning between CP and TX, since typically CP and TX data are available for compounds in our library. However, for new compounds, it is highly likely that only CP data is available, as TX data is comparatively more costly to obtain.

## 2 Methods

For feature learning, we train two methods: Contrastive Learning and Bimodal Autoencoder. Contrastive Learning [1] learns representations using one Multi-layer Perceptron (MLP) encoder for each modality. The model is trained by minimizing a contrastive loss that forces the embeddings of the same compound close, and different compounds far away. Then embeddings are extracted from the CP encoder. Bimodal Autoencoder [2] learns representation using one MLP encoders for CP, and two MLPs decoder for CP and TX. The model learns to

© The Author(s) 2025
D.-A. Clevert et al. (Eds.): AIDD 2024, LNCS 14894, pp. 169–170, 2025.
https://doi.org/10.1007/978-3-031-72381-0

reconstruct both modalities using only CP. The embeddings are then extracted from the middle layer.

To understand how good the embeddings are in which tasks, compared to directly using CP or TX as features, we benchmark them on a range of activity modelling downstream tasks.

## 3   Results

Preliminary results show on average across all tasks, performances of models using the embeddings are comparable to using CP or TX features. However, the embeddings manage to improve performances in underperforming tasks that only use one modality. In terms of unsupervised tasks, embeddings improve clustering based on cell painting replicates and modes of action.

**Acknowledgement.** This study was partially funded by the European Union's Horizon 2020 research and innovation programme under the Marie Skłodowska-Curie Actions grant agreement "Advanced machine learning for Innovative Drug Discovery (AIDD)" No. 956832.

## References

1. Radford, A., et al.: Learning transferable visual models from natural language supervision. CoRR, abs/2103.00020 (2021). https://arxiv.org/abs/2103.00020
2. Ngiam, J., et al.: Multimodal deep learning. In: Proceedings of the 28th International Conference on Machine Learning, ICML 2011, Bellevue, Washington, USA, 28 June – 2 July 2011, pp. 689–696. Omnipress (2011)
3. Baltrusaitis, T., Ahuja, C., Morency, L.P.: Multimodal machine learning: a survey and taxonomy. IEEE Trans. Pattern Anal. Mach. Intell. **41**(2), 423–443 (2019)
4. Sanchez-Fernandez, G.: CLOOME: contrastive learning unlocks bioimaging databases for queries with chemical structures. Nat. Commun. **14**(1), 7339 (2023). https://doi.org/10.1038/s41467-023-42328-w

# Accelerating the Inference of String Generation-Based Chemical Reaction Models for Industrial Applications

Mikhail Andronov[1,2(✉)] [ID], Natalia Andronova[5], Michael Wand[1,3], Jürgen Schmidhuber[1,4] [ID], and Djork-Arné Clevert[2] [ID]

[1] IDSIA, USI, SUPSI, 6900 Lugano, Switzerland
{mikhail.andronov,michael.wand}@idsia.ch
[2] Machine Learning Research, Pfizer Research and Development, Friedrichstr. 110, 10117 Berlin, Germany
djork-arne.clevert@pfizer.com
[3] Institute for Digital Technologies for Personalized Healthcare, SUPSI, 6900 Lugano, Switzerland
[4] AI Initiative, KAUST, Thuwal 23955, Saudi Arabia
juergen.schmidhuber@kaust.edu.sa
[5] Lugano, Switzerland

**Abstract.** Automated planning of organic chemical synthesis, first formalized around fifty years ago [6], is one of the core technologies enabling computer-aided drug discovery. While first computer-aided synthesis planning (CASP) systems relied on manually encoded rules [2, 4], researchers now primarily focus on CASP methods powered by artificial intelligence techniques. The design principles of the latter were outlined in the seminal work by Segler et al. [9]: a machine learning-based single-step retrosynthesis model combined with a planning algorithm. The former proposes several candidate retrosynthetic chemical transformations for a given molecule, and the latter, e.g., Monte-Carlo Tree Search, uses those candidates to construct a synthesis tree. One of the way of doing single-step retrosynthesis is to pose the task as SMILES-to-SMILES translation. Both reaction product prediction and single-step retrosynthesis can be done in this way. The transformer [7, 12], a neural architecture initially proposed for neural machine translation, adapts well to SMILES-to-SMILES translation and now serves as a backbone for state-of-the-art models for single-step retrosynthesis [3, 10] and reaction product prediction [3, 8].

As transformer-based single-step retrosynthesis models reach state-of-the-art accuracy in the task, they become increasingly interesting for integration in multi-step synthesis planning systems. However, the latency of the transformer hinders its suitability for synthesis planning [11]. Therefore, a need for accelerating the inference of the SMILES-to-SMILES transformer arises.

Recently, transformer-based Large Language Models (LLMs) [1] gained a lot of attention and attracted a lot of research effort. One of the

N. Andronova—Independent Researcher

© The Author(s) 2025
D.-A. Clevert et al. (Eds.): AIDD 2024, LNCS 14894, pp. 171–173, 2025.
https://doi.org/10.1007/978-3-031-72381-0

critical issues in LLMs is their slow inference speed. Speculative decoding [5] is a technique that allows to generate more than one token for one forward pass of the model.

In our work, we propose a method to accelerate inference from SMILES-to-SMILES translation models based on speculative decoding combined with insights from the chemical essence of the problem. We reimplement the Molecular Transformer [8] in Pytorch Lightning and use our method to demonstrate its inference acceleration in single-step retrosynthesis and product prediction by three times without changing the model architecture or the training procedure. Our implementation of the Molecular Transformer (MT) successfully reproduces the accuracy scores of the original MT [8] that relies on OpenNMT. Comparing our MT and the original MT, we observe at most 0.2% points discrepancy of top-1 to top-5 accuracy in product prediction with beam search. We use are approach to significantly accelerate greedy decoding from MT in reaction product prediction on USPTO MIT and beam search in single-step retrosynthesis on augmented USPTO 50K.

**Acknowledgments.** This study was partially funded by the European Union's Horizon 2020 research and innovation program under the Marie Skłodowska-Curie Innovative Training Network European Industrial Doctorate grant agreement No. 956832 "Advanced machine learning for Innovative Drug Discovery", and also by the Horizon Europe funding programme under the Marie Skłodowska-Curie Actions Doctoral Networks grant agreement "Explainable AI for Molecules - AiChemist" No. 101120466.

# References

1. Brown, T., et al.: Language models are few-shot learners. Adv. Neural. Inf. Process. Syst. **33**, 1877–1901 (2020)
2. Gasteiger, J., et al.: Computer-assisted synthesis and reaction planning in combinatorial chemistry. Perspect. Drug Discov. Des. **20**, 245–264 (2000)
3. Irwin, R., Dimitriadis, S., He, J., Bjerrum, E.J.: Chemformer: a pre-trained transformer for computational chemistry. Mach. Learn. Sci. Technol. **3**(1), 015022 (2022)
4. Johnson, P.Y., Burnstein, I., Crary, J., Evans, M., Wang, T.: Designing an Expert System for Organic Synthesis: the Need for Strategic Planning. ACS Publications, Washington, DC, USA (1989)
5. Leviathan, Y., Kalman, M., Matias, Y.: Fast inference from transformers via speculative decoding. In: International Conference on Machine Learning, pp. 19274–19286. PMLR (2023)
6. Pensak, D.A., Corey, E.J.: LHASA—Logic and Heuristics Applied to Synthetic Analysis. ACS Publications, Washington, DC, USA (1977)
7. Schmidhuber, J.: Learning to control fast-weight memories: an alternative to recurrent nets. Neural Comput. **4**(1), 131–139 (1992)
8. Schwaller, P., et al.: Molecular transformer: a model for uncertainty-calibrated chemical reaction prediction. ACS Central Sci. **5**(9), 1572–1583 (2019)
9. Segler, M.H., Preuss, M., Waller, M.P.: Planning chemical syntheses with deep neural networks and symbolic AI. Nature **555**(7698), 604–610 (2018)

10. Tetko, I.V., Karpov, P., Van Deursen, R., Godin, G.: State-of-the-art augmented NLP transformer models for direct and single-step retrosynthesis. Nat. Commun. **11**(1), 5575 (2020). https://doi.org/10.1038/s41467-020-19266-y
11. Torren-Peraire, P., et al.: Models matter: the impact of single-step retrosynthesis on synthesis planning. Digit. Discov. **3**(3), 558–572 (2024)
12. Vaswani, A., et al.: Attention is all you need. Adv. Neural Inf. Process. Syst. **30** 5999–6010 (2017)

# Author Index

© The Editor(s) (if applicable) and The Author(s) 2025
D.-A. Clevert et al. (Eds.): AIDD 2024, LNCS 14894, pp. 175–176, 2025.
https://doi.org/10.1007/978-3-031-72381-0